1 MONTH OF FREE READING

at

www.ForgottenBooks.com

By purchasing this book you are eligible for one month membership to ForgottenBooks.com, giving you unlimited access to our entire collection of over 1,000,000 titles via our web site and mobile apps.

To claim your free month visit: www.forgottenbooks.com/free899778

* Offer is valid for 45 days from date of purchase. Terms and conditions apply.

ISBN 978-0-265-85495-2
PIBN 10899778

This book is a reproduction of an important historical work. Forgotten Books uses state-of-the-art technology to digitally reconstruct the work, preserving the original format whilst repairing imperfections present in the aged copy. In rare cases, an imperfection in the original, such as a blemish or missing page, may be replicated in our edition. We do, however, repair the vast majority of imperfections successfully; any imperfections that remain are intentionally left to preserve the state of such historical works.

Forgotten Books is a registered trademark of FB &c Ltd.
Copyright © 2018 FB &c Ltd.
FB &c Ltd, Dalton House, 60 Windsor Avenue, London, SW19 2RR.
Company number 08720141. Registered in England and Wales.

For support please visit www.forgottenbooks.com

PEABODY MUSEUM OF NATURAL HISTORY, YALE UNIVERSITY
BULLETIN 10

Revision of the Gymnarthridae American Permian Microsaurs

BY

JOSEPH T. GREGORY,
FRANK E. PEABODY,
and LLEWELLYN I. PRICE

NEW HAVEN, CONNECTICUT

1956

Printed in the United States of America

CONTENTS

ABSTRACT	ix
INTRODUCTION	1
THE FORT SILL LOCALITY	2
ACKNOWLEDGMENTS	3
ABBREVIATIONS	4
REVISION OF THE MICROSAUR FAMILY GYMNARTHRIDAE	5
Pariotichus brachyops Cope	6
Species incorrectly referred to *Pariotichus*	8
Cardiocephalus sternbergi Broili	9
Cardiocephalus cf. *sternbergi* Broili from Oklahoma	14
Y.P.M. no. 3689, skull and jaws	15
K.U.M.N.H. no. 8967, skull and jaws	19
O.U. no. 1034, articulated skeleton	24
Euryodus primus Olson	26
Pantylus, Ostodolepis, and *Goniocara*	29
P. cordatus Cope	30
P. coicodus Cope	31
P. tryptichus Cummins	31
O. brevispinatus Williston	31
G. willistoni Broili	32
Systematic summary	32
Stratigraphic distribution of the Gymnarthridae	34
THE AUDITORY REGION OF MICROSAURS	35
GYMNARTHRID TOOTH IMPLANTATION AND SUCCESSION	38
ISOLATED POSTCRANIAL BONES OF GYMNARTHRIDS	41
Occipito-atlas region	41
Vertebrae	44
Ribs	49
Shoulder girdle	50
Pelvis	52
Limb bones	55
RELATIONSHIPS OF THE MICROSAURS	63
RESTORATION	69
REFERENCES CITED	71
INDEX	75

ILLUSTRATIONS

Fort Sill fossil locality, and restoration of *Cardiocephalus*   Frontispiece

Figure 1.	*Pariotichus brachyops* Cope, type skull	6
Figure 2.	*Pariotichus brachyops* Cope, restoration of skull roof	7
Figure 3.	*Cardiocephalus sternbergi* Broili, type skull and restoration	10
Figure 4.	*Cardiocephalus sternbergi* Broili. (Type skull of *Gymnarthrus willoughbyi* Case)	12
Figure 5.	*Cardiocephalus sternbergi* Broili. (Paratype of *Gymnarthrus*)	12
Figure 6.	*Cardiocephalus* cf. *sternbergi* Broili, skull	15
Figure 7.	*Cardiocephalus* cf. *sternbergi* Broili, lower jaws	17
Figure 8.	*Cardiocephalus* cf. *sternbergi* Broili, skull and jaws	20
Figure 9.	*Cardiocephalus* cf. *sternbergi* Broili, articulated skeleton	24
Figure 10.	*Cardiocephalus* cf. *sternbergi* Broili, restoration of skull	25
Figure 11.	*Euryodus primus* Olson, type skull	27
Figure 12.	*Euryodus primus* Olson, dentition	30
Figure 13.	*Cardiocephalus sternbergi* Broili, otic region	36
Figure 14.	*Euryodus primus* Olson, tooth replacement	38
Figure 15.	*Euryodus primus* Olson, sections of teeth	39
Figure 16.	Atlas, basioccipital—exoccipital complex, and basisphenoid—parasphenoid complex of gymnarthrids	42
Figure 17.	*Cardiocephalus*, dorsal vertebra	44
Figure 18.	*Cardiocephalus*, vertebrae	45

Figure 19.	*Cardiocephalus* AND *Euryodus*, VERTEBRAE	46
Figure 20.	?*Euryodus* OR ?*Captorhinus*, VERTEBRAE AND *Captorhinus* VERTEBRA	48
Figure 21.	?*Cardiocephalus*, SCAPULOCORACOID AND PELVIS	50
Figure 22.	*Cardiocephalus*, SCAPULOCORACOID AND ISCHIUM	51
Figure 23.	*Cardiocephalus*, PELVIS	53
Figure 24.	*Cardiocephalus*, HUMERUS	55
Figure 25.	*Cardiocephalus*, HUMERI	56
Figure 26.	*Captorhinus*, HUMERUS	56
Figure 27.	?*Euryodus*, ULNA AND *Cardiocephalus*, ULNAE AND RADII	57
Figure 28.	*Cardiocephalus*, FEMORA	58
Figure 29.	*Captorhinus*, FEMUR, TIBIA, AND FIBULA	59
Figure 30.	*Cardiocephalus*, TIBIAE AND FIBULAE	60
Figure 31.	INDEX TO MEASUREMENTS OF LIMB BONES	61
Figure 32.	HISTOGRAMS OF VARIATION IN LENGTH OF MICROSAUR LIMB BONES FROM FISSURE NORTH OF FORT SILL, OKLAHOMA	62
Figure 33.	SUGGESTED PHYLETIC RELATIONSHIPS OF LEPOSPONDYL ORDERS	68

TABLES

Table 1.	Comparative skull measurements	14
Table 2.	Tooth counts of *Cardiocephalus*	18
Table 3.	Tooth counts of *Euryodus* specimens	29
Table 4.	Stratigraphic distribution of gymnarthrid genera	34
Table 5.	Measurements in millimeters of vertebrae	49
Table 6.	Measurements in millimeters of scapulocoracoid	52
Table 7.	Measurements in millimeters of limb bones	61

ABSTRACT

Pariotichus is known from a single damaged skull, the type of *Pariotichus brachyops* Cope; all other specimens which have been referred to the genus are captorhinomorph reptiles. The type specimens of *Gymnarthrus* are conspecific with *Cardiocephalus sternbergi* Broili; abundant material from fissures in a limestone quarry north of Fort Sill, Oklahoma, probably referable to the same species, shows many details of skull and postcranial anatomy. *Euryodus primus* Olson differs from *Cardiocephalus* in larger size, small nondentigerous coronoids in lower jaw and round rather than compressed crushing teeth, one of which is extremely enlarged. *Pantylus, Ostodolepis,* and possibly *Goniocara* are related genera with more crushing, less trenchant adaptations of the dentition.

The stapes of *Cardiocephalus* has a large footplate and short imperforate columella resembling that of urodeles and aistopods. Skull structures do not permit definite determination of the relationships of microsaurs among lepospondylous amphibians. The transverse process on the dorsal vertebrae of microsaurs is at the anterior end as in gymnophionans and lysorophids, in contrast to the transverse processes of nectridians, aistopods, and urodeles, which arise near the middle of the vertebrae.

Phylogenetically this could mean that the gymnophionans rather than urodeles are descendants of the microsaurs, and that urodeles may be an offshoot of primitive nectridians.

REVISION OF THE GYMNARTHRIDAE
AMERICAN PERMIAN MICROSAURS

INTRODUCTION

Small size and generally poor preservation of the Pennsylvanian and early Permian microsaurs have prevented a clear understanding of their anatomy and led to widely divergent views regarding their phylogenetic position. Many problems regarding the content and nature of this Order were solved by Romer (1950). Discovery of far better preserved specimens than had previously been known affords an opportunity to confirm and extend his systematic conclusions, and at the same time to revise in detail the family of American Permian microsaurs currently called Gymnarthridae.

In 1932 operators of the Dolese Brothers limestone quarry between Fort Sill and Apache, Commanche County, Oklahoma, brought to the attention of geologists at the University of Oklahoma masses of tiny bones imbedded in clay which filled fissures in the limestone. Extensive collections were made by students, including Llewellyn I. Price. In addition to innumerable isolated bones, mainly belonging to the small cotylosaur *Captorhinus*, a nodule was found which contained an articulated skeleton of the microsaur *Cardiocephalus*. Later, at Harvard University, Price prepared drawings of the skull of this specimen and also assembled and drew all other known gymnarthrids save the types of *Cardiocephalus*, which are in Germany. He likewise assembled notes for a revision of the family, but was delayed in publishing them by his departure for Brazil. His unpublished conclusions were utilized by Professor Romer in preparing the discussion of microsaurs in the 1945 revision of "Vertebrate Paleontology."

An unusually well preserved microsaur skull was obtained from the same Permian fissure deposits north of Fort Sill, by the 1947 field party of Peabody Museum. The specimen was brought back from the field undetected in a block of matrix and was not exposed during the preliminary washing of the material. During the fall of 1950, Mr. G. Donald Guadagni, then chief preparator of Vertebrate Paleontology in Peabody Museum, discovered the skull while working out the more resistant nodules. He removed the tenacious incrustation of calcite and pyrite from the tiny skull with needles and dental burrs, working under a binocular microscope. The perfection of the specimen is a tribute to Mr. Guadagni's skill and patience at delicate preparation.

Unaware of Price's work, Joseph T. Gregory commenced a revision of the gymnarthrids, had illustrations prepared of the various types, and arrived at substantially the same conclusions which Price had reached a decade before.

Meanwhile in 1949 and 1950, Frank E. Peabody collected from this locality for the University of Kansas Museum of Natural History. He discovered remains of *Euryodus* and, in 1950, a *Cardiocephalus* skull with jaws in place which revealed new morphological data. He began a study which included, through the kindness of Drs. Everett Olson and Rainer Zangerl, materials collected by the Chicago Museum of Natural History. At the 1951 meeting of the Society of Vertebrate Paleontology in Detroit the authors met and agreed to pool their data for a comprehensive revision of the family.

It was decided that the task of assembling and coordinating these studies would be carried out by Gregory, whose manuscript was most advanced at the time. During the first half of 1952 the descriptive and systematic portions were completed, except for description of postcranial elements, whose identification and separation from rather similar bones of *Captorhinus* occupied most of that summer. That fall other work intervened, and the project stood still until the fall of 1954 when study of the limb bones and vertebrae was completed and the remaining sections were written.

Whether the larger animals *Pantylus* and *Ostodolepis* belong within the Gymnarthridae or form an allied family, the Pantylidae, is unimportant. They have been included in the systematic summary given below, but no attempt is made to evaluate their affinities critically, pending the results of a detailed study of important new specimens of *Pantylus* by A. S. Romer and J. A. Wilson.

Examination of the type of *Isodectes megalops* (Cope) in the American Museum confirms its assignment to the captorhinid cotylosaurs, but it is possible that the skull referred to this species by Williston (1916, p. 176–178, fig. 32, W.M. no. 686) is actually a gymnarthrid. It cannot be located at present, and has not been included in the present study.

THE FORT SILL LOCALITY

New material described in this paper came from Dolese Brothers limestone quarry, situated just west of the Anadarko–Lawton highway (U.S. routes 62 and 281), 10¼ miles south of Apache, about 1¼ miles south of the junction of this route and highway 277 (from Chickasha). It is about 6 miles north of Fort Sill, in sec. 31, T. 4 N. R. 11 W., Indian Meridian and Baseline, in Comanche County, Oklahoma. The quarry is in steeply dipping beds (60° E.) of Arbuckle limestone at the eastern end of the Wichita Mountains uplift. Near the north end of the old east-facing quarry are at least three almost vertical fissures (Frontispiece) filled with limestone breccia and pockets of soft blue and yellow clay, all containing myriads of bones of small reptiles and amphibians. These fissures vary from a few centimeters to perhaps a meter in width and have irregularly curved walls whose pitted surface is typical of the texture produced on limestones by solution. Their lower limit is not exposed; above, they open into the soil zone of the hilltop, a zone containing scattered bones of recent or subrecent age. At present, quarry work is advancing from the south, at right angles to the old working face and will eventually destroy all evidence of the three fissures described above. The new working, however, continues to expose new deposits of blue clay containing countless bones similar to those in the old quarry. Hereafter this locality will be termed the "Fort Sill Locality," or "Fissures north of Fort Sill." Olson (1954, p. 211) refers to the same spot as "Richard's Spur, Oklahoma."

Cardiocephalus is represented among specimens from these fissures by an articulated skeleton (O.U. no. 1034), two skulls (Y.P.M. no. 3689, K.U.M.N.H. no. 8967), numerous jaws, and hundreds of isolated vertebrae and limb bones. Apparently it was second only to *Captorhinus* in abundance in the fissures. Other members of the assemblage are *Euryodus*, rare *Labidosaurus*, a carnivorous cotylosaur cf. *Romeria*, small ophiacodont pelycosaurs, medium sized pelycosaurs of uncertain family, infrequent small labyrinthodonts, and an aistopod (recognized from a single vertebra). The assemblage is similar to that of the Arroyo fauna (Lower Clear Fork) of Texas, although of different facies; most species appear identical to those of the Arroyo.

ACKNOWLEDGMENTS

The writers are grateful to Mr. J. A. Yingling, superintendent of the Dolese Brothers Limestone Quarry at Apache, Oklahoma, for permission to collect at this locality. Dr. E. H. Colbert and Mrs. Rachel H. Nichols of the American Museum of Natural History have most kindly allowed us to compare the Oklahoma material with type material from Texas, and to refigure the latter. Opportunity to study and further prepare an articulated skeleton of *Cardiocephalus* has been afforded by the late Profesor J. Willis Stovall of the University of Oklahoma and Professor A. S. Romer of Harvard University. We are also indebted to Professor Romer for critical comments upon the manuscript. Drs. E. C. Olson and Rainer Zangerl of the Chicago Natural History Museum generously turned over a large collection of material from the Fort Sill locality for study, and also have permitted restudy of specimens described by Williston. Thanks are likewise due to Mr. William Turnbull of that Museum for assistance with this material. Professor C. L. Camp of the University of California has furnished notes on the types of *Cardiocephalus* in Munich. Mr. C. M. Bogert has permitted study of recent caecelian vertebrae in the herpetology collection of the American Museum, and Mr. Sam MacDowell has assisted in cleaning this material for study and discussed problems of lepospondyl relationships. Professor Carl O. Dunbar of Yale University gave access to the unpublished National Research Council Correlation Chart of Permian Rocks of North America.

Illustrations, unless otherwise acknowledged, have been prepared by Miss Shirley Glaser, staff artist at Peabody Museum. Drawings of specimens at the University of Kansas are by Mr. Victor Hogg, or by Frank E. Peabody with the assistance of Mr. Hogg. Some illustrations of the skulls are by Llewellyn I. Price. Special acknowledgment is due the late Mrs. Mildred

Porter Cloud, librarian at Peabody Museum, for painstaking editorial assistance.

ABBREVIATIONS

The following abbreviations are used throughout the text to refer to museum collections:

A.M.N.H.	American Museum of Natural History, New York City, N.Y.
C.N.H.M.	Chicago Natural History Museum (formerly Field Museum), Chicago, Illinois.
K.U.M.N.H.	Kansas University Museum of Natural History, Lawrence, Kansas.
O.U.	University of Oklahoma Museum, Norman, Oklahoma.
U.M.M.P.	University of Michigan, Museum of Paleontology, Ann Arbor, Michigan.
W.M.	Walker Museum collection of University of Chicago, now in Chicago Natural History Museum, Chicago, Illinois.
Y.P.M.	Peabody Museum of Natural History, Yale University, New Haven, Connecticut.

REVISION OF THE MICROSAUR FAMILY GYMNARTHRIDAE

E. C. Case described *Gymnarthrus* in 1910 and erected a new family, Gymnarthridae, and suborder, Gymnarthria, for it. At first he regarded it as a reptile, but the next year (1911b, p. 14, 69, 145) following Broom (1910, p. 220) he placed it *incertae sedis* among the Stegocephalia, and included *Cardiocephalus* Broili, 1904, in the same family. Gymnarthridae is now generally used for these small amphibians (cf. Romer, 1950, p. 638–639).

Many years earlier E. D. Cope established the family Pariotichidae (1883, p. 631) for the genera "*Pariotichus, Pantylus,* and probably *Ectocynodon.*" He defined the family as having teeth like the Edaphosauridae (more than one series of teeth in the jaws) but differing from it in the entire roof over the temporal fenestra. *Pariotichus,* the type genus of the family, had been described (Cope, 1878, p. 502) from a single crushed and incomplete skull, *P. brachyops* Cope. In later articles Cope (e.g. 1895, p. 442–452) sought to supply missing details of the structure of *Pariotichus* from other species which he referred to the genus. Now, it is apparent that almost from the beginning of his studies of Permian vertebrates, he regarded the small *Pariotichus brachyops* skull as belonging to the group of cotylosaurs which today is termed the Captorhinidae, and that he referred various captorhinomorph reptiles to the genus *Pariotichus,* gradually building up a definition and concept of the genus and family based upon these referred materials rather than the original type.

Case (1911a, p. 33–49) recognized Cope's error and separated the family Captorhinidae from the Pariotichidae. However like Cope, he drew largely upon a referred "homotype" skull (A.M.N.H. 4760) for his definition of the genus *Pariotichus.* Broom (1930, p. 48–50) recognized that this skull also differed from *Pariotichus brachyops* and actually belonged to *Captorhinus,* so he made it the type of species *C. gregoryi* Broom.

In the original description of *P. brachyops,* Cope stated that the skull had lost its outer layer of bone. Careful inspection by Price in 1938 revealed that an incrustation of matrix covered the skull roof; he carefully removed this, revealing the crushed remnants of the roofing bones, not of a captorhinomorph reptile but of a microsaur similar to *Cardiocephalus* or *Euryodus.* Romer (1945, p. 592; 1950, p. 639) referred *Pariotichus* to the microsaur family Gymnarthridae on the basis of these unpublished studies by Price. Accordingly a redescription of this specimen is desirable, the more so since no other material properly referrable to *Pariotichus* has been discovered.

PARIOTICHUS BRACHYOPS COPE (1878, p. 502)

TYPE: A.M.N.H. no. 4328, imperfect skull.
LOCALITY: North Fork Little Wichita River, Archer Co., Texas.
AGE: Either Putnam or Admiral Formation, Wichita Group, Lower Permian. Zone O or I of Romer (1928).

The skull was collected by Jacob Boll in 1878 from the lower portion of the valley of the North Fork of the Little Wichita River. Romer (1928, p. 77) has shown that Boll's collection that year was entirely from the lower part of the fossiliferous redbed section. Cope's original description states that *P. brachyops* was from the same locality as *Bolosaurus rapidens* (= *Chilonyx rapidens*, A.M.N.H. no. 4356).

REDESCRIPTION OF TYPE: The type, and only known specimen, is a crushed skull lacking the occiput and tip of the snout (fig. 1). The parietals, inter-

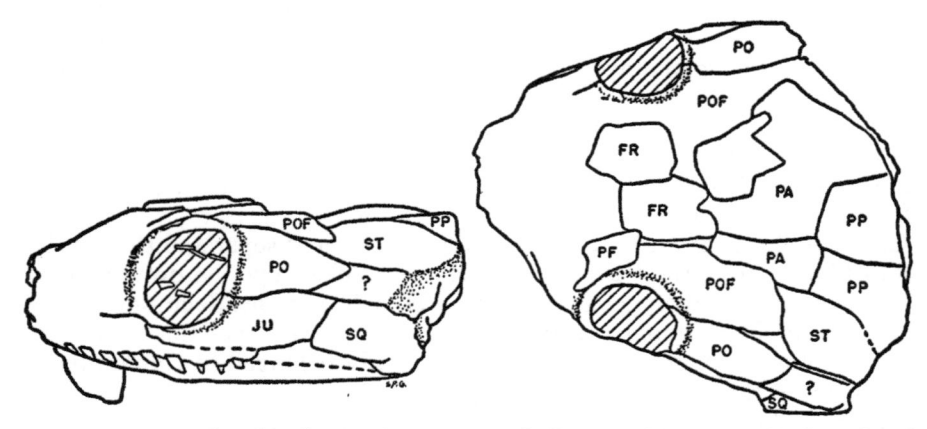

Figure 1. *Pariotichus brachyops* Cope. Type skull, A.M.N.H. no. 4328, dorsal and left lateral views. x 3. FR frontal, PA parietal, PF prefrontal, POF postfrontal, PP postparietal, PO postorbital, SQ squamosal, ST supratemporal.

parietals and right supratemporals, as a unit, have been pushed forward and to the left underneath the frontals and left side of the skull. The elements of the left orbital region and the supratemporal have preserved their proper relations. The left prefrontal has shifted forward slightly, but that of the right side has retained its normal position with postfrontal, frontal, nasal, and lacrimal. The anterior surface of the snout is damaged but indicates, very nearly, the anterior extent of the face. Some uncertainty exists as to the length of parietals and frontals, and the anterior bones are obscured. Nevertheless essential features of the arrangement of the dermal bones are clear and restorations prepared independently by Price and Gregory agree in most respects.

The skull is flat, triangular, broad across the occiput, with a probably blunt, rounded snout. Its length as preserved is 21 mm., and it probably was about 25 mm. long by 20 mm. wide in life. Small circular orbits (4½ mm. in diameter) lie forward of the middle of the head. The orbital borders turn outward to form a raised rim which is separated from the skull roof at least dorsally by a shallow groove. Anterior nares cannot be determined but must have been small and far forward on the snout.

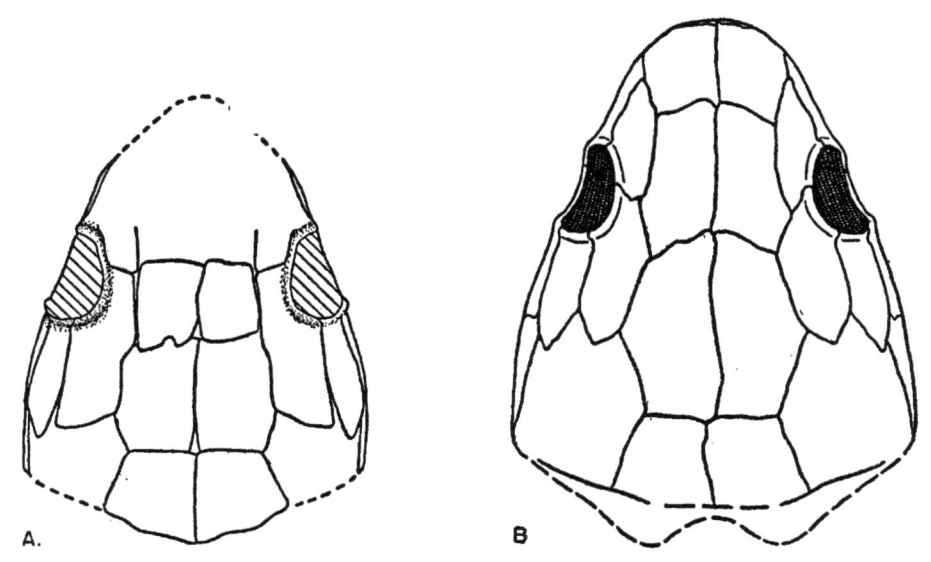

Figure 2. *Pariotichus brachyops* Cope. Restoration of skull roof. A. by Shirley P. Glaser; B. by L. I. Price. Both x 3.

Important features of the roofing bone pattern (shown in fig. 2) are the large postparietal, a very large supratemporal reaching the rear of the skull roof, absence of a tabular, and small laterally placed squamosal and quadratojugal. The pattern is clearly that regarded as typical of microsaurs by Romer (1950, fig. 1).

The surface of the roofing bones is smooth.

The shattered palate shows no structure except a broad plate of bone (parasphenoid) beneath the brain case. The jaws lie in place, largely concealed from view by the maxillaries.

Eleven upper teeth can be distinguished on the right side of the skull. These all appear to belong to the maxillary, and it cannot be determined whether other maxillary teeth preceded them, nor how many premaxillary teeth there were. The teeth are short, with blunt conical tips on some of which longitudinal striations similar to those of the better known *Cardiocephalus* and *Euryodus* teeth may be seen. Tooth diameter is greatest in the region just in front of the orbit, but there is no greatly enlarged tooth like that of *Euryodus*. On both sides the three posterior teeth are abruptly smaller than those in front of them.

These meager details indicate that our knowledge of *Pariotichus brachyops* is most unsatisfactory. Its relationship to the gymnarthrid microsaurs is shown by the pattern of the posterior roofing bones of the skull and by the form and arrangement of the maxillary teeth. This species comes from the Wichita group and thus is older than its relatives *Cardiocephalus*, *Euryodus*, etc., all of which are known only from the Clear Fork group. Its size is intermediate between *Cardiocephalus* and *Euryodus*, and its skull appears to differ slightly in proportions, being relatively wider in proportion to its length. In view of the possible significance of these differences, the earlier age, and the poor preservation of the type and only known specimen, it seems best to regard the genus as distinct from the Clear Fork gymnarthrids. Should more material be found in these lower beds, it may be possible to define the genus more precisely.

SPECIES INCORRECTLY REFERRED TO PARIOTICHUS

Pariotichus megalops Cope 1883 = *Isodectes megalops* (Cope)
Type of *Isodectes*, Cope 1895

Pariotichus aduncus Cope 1896 = *Captorhinus aduncus* (Cope), Case 1911

Pariotichus isolomus Cope 1896 = *Captorhinus isolomus* (Cope), Case 1911

Pariotichus laticeps Williston 1909 = *Captorhinus*—not included in Case's revision

Pariotichus hamatus Cope 1895 = *Labidosaurus hamatus* (Cope) 1896

Ectocynodon aguti Cope 1882 = *Captorhinus aguti* (Cope)
(By Cope 1895)

Ectocynodon incisivus Cope 1886 = *Captorhinus aguti* (Cope), Case 1911
(By Cope 1895)

Ectocynodon ordinatus Cope 1878 = Type of *Ectocynodon*, a captorhinid
(By Cope 1895)

Captorhinus angusticeps Cope 1895 (Type of *Captorhinus*)
(By Broom 1910)

As Williston (1911b, p. 68) pointed out, all the species listed above are captorhinomorph reptiles; most were properly assigned to *Captorhinus* by Case in his 1911 revision of the cotylosaurs. Further consideration of these species is not germane to the present problem.

CARDIOCEPHALUS STERNBERGI BROILI (1904, p. 45)

SYNONYM: *Gymnarthrus willoughbyi* Case (1910, p. 177).
TYPES: Holotype skull and referred skull and vertebrae, Alte Akademie, Munich, Germany.
LOCALITY: Coffee Creek Bonebed, Baylor County, Texas.
AGE: Arroyo formation (Clear Fork Group, L. Permian) about 100 feet above base. Zone 4 of Romer (1928).

A.M.N.H. no. 4892 (Type of *Gymnarthrus willoughbyi* Case). The label on this specimen reads Baylor County; Case in the original description (1910, p. 177) states that it is "From Baylor Co. near the head of Coffee Creek, in a red clay above the Wichita Conglomerate." In 1911 (p. 69) however, Case stated that it was from Wilbarger County and Romer, 1928, p. 83–85, states that the type was from Beaver Creek, Wilbarger County from beds above the Leuders limestone and probably corresponding to those on Coffee Creek. In spite of this confusion there is no suggestion that the specimen was found other than in an area of Clear Fork beds.

A.M.N.H. no. 4763a (Paratype of *Gymnarthrus willoughbyi* Case). Locality unknown, from the Cope Collection according to Case; L. I. Price's notes state that it was collected by C. H. Sternberg in 1902 from Hackberry Creek, one of the localities in the valley of the Little Wichita southeast of Fulda which Romer places in his Zone 2, Belle Plains Formation, middle Wichita Group. The close similarity to the type specimen, and the lack of confirmation of the locality, cause me to question this early occurrence of the genus.

W.M. no. 1047. Referred skull.
Beaver Creek near Vernon Road, Wilbarger County, Texas. Collected by Paul Miller, 1914. Arroyo formation. This would be essentially a topotype of *"Gymnarthrus"* if Case's second statement is correct. However, it is possible that Romer's reference of the type of *Gymnarthrus* to the Beaver Creek locality was actually based on this specimen.

DISCUSSION OF TYPES: In 1904 Broili described two small skulls from the Coffee Creek bonebed (Arroyo formation, Clear Fork group) of Texas. His excellent figure (Broili, 1904, Pl. 6, figs. 5–5a) shows the wide temporal region, forwardly placed orbits, well developed postparietals, and large supratemporal of the microsaurs. The circumorbital groove which characterizes most gymnarthrid specimens is well marked, especially on the prefrontal, and was interpreted by Broili as a slime canal ("Lyra") remnant of the lateral line sensory system. He observed 10 teeth, small anteriorly, enlarged below the orbit, compressed and bearing anterior and posterior ridges.

It has not been possible for us to examine the cotypes, which are in the Alte Akademie in Munich. In 1935 Professor C. L. Camp examined the original specimens and made the following notes, which he has kindly permitted us to use: "tiny, small-eyed, flat-headed form, with protruding muzzle, small underslung nostrils and hard porcelain-like plates over the head. Perhaps a burrowing form . . . no indication of a parietal (pineal) foramen . . . stoutly built with enormous lacrymals and prefrontals. Squamosal extensive and overhangs the vault for the jaw muscles. The smooth, polished surface of the bones is interesting . . . must have been covered with thin, polished scales . . . apparently an osteodermal throat plate— of polished scales, and the entire gular region *was* covered with osteoderms which have been torn away during preparation [fig. 3B] . . . The palate is

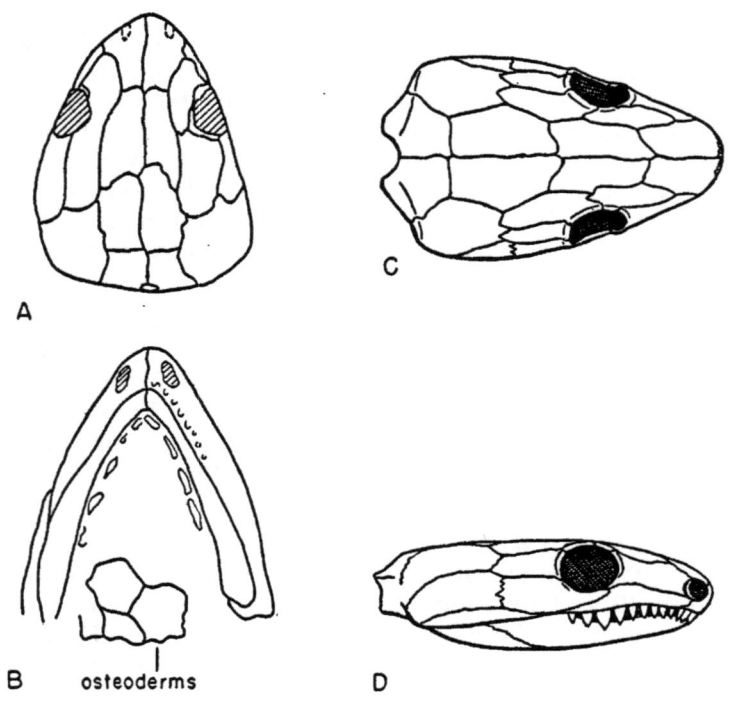

Figure 3. *Cardiocephalus sternbergi* Broili. A. dorsal and B. ventral views of one of cotype skulls, Alte Akademie, Munich, from sketch by C. L. Camp. x 3. C. and D. restoration of skull by L. I. Price based upon paratype of *Gymnarthrus* and Broili's figures. Dorsal and lateral views. x 3.

exceptionally flat and massive—amphibian-like. Teeth resemble *Captorhinus*—those at front being much smaller than those on mx. and all are short, blunt and very smooth. This is Case's *Gymnarthrus*." One may suspect that the small teeth which generally occur behind the enlarged maxillary teeth of these amphibians were concealed or lost from Broili's specimens. The size, form, and proportions of the skull, and arrangement of roofing bones, agree in all respects with the well preserved skull from Oklahoma described in this paper. Camp does not mention the raised orbital rims, but this feature seems variable among other specimens which we have examined, and Broili's illustrations and description clearly indicate the existence of a circumorbital groove.

Broili was in doubt as to the affinities of this small animal as his material did not show the palate or occiput. He referred it to the Stegocephalia because of the supposed slime canals. Case (1911b, p. 145) recognized the close affinity of *Cardiocephalus* and *Gymnarthrus*, and concurred with Broom (1910, p. 219–220) as to the amphibian affinities of these forms.

In view of the unsatisfactory type of *Pariotichus*, it seems unwise to synonymize *Cardiocephalus* with the Wichita genus, although morphological differences are not recognizable at present. Most members of the Clear Fork fauna have been found specifically or even generically distinct from their Wichita counterparts when adequate material was available for diagnosis.

To *Cardiocephalus sternbergi* are referred the types and referred specimens of *Gymnarthrus willoughbyi* Case from the Arroyo formation of Texas. The smaller gymnarthrid skulls and jaws from the Fort Sill, Oklahoma, fissures also are very close to this species.

Gymnarthrus willoughbyi Case was based upon two skulls from the lower Clear Fork group of Texas (figs. 4–5). Both are broken in the temporal region in such a way that a turtle-like excavation of the cheek behind the maxilla is suggested. In the original description Case (1910, p. 178–179) recognized that the closest affinities were with *Cardiocephalus*, from which he distinguished *Gymnarthrus* by:

1. Temporal region excavated below, quadratojugal lost and (pro) squamosal reduced.
2. Parietal foramen present.
3. No "lyra."
4. No cutting edges on teeth, the last maxillary tooth small.

In 1911 (p. 14), after direct comparison with the type of *Cardiocephalus*, he reiterated these features but emphasized the so-called "lyrae" and parietal foramen as generic distinctions. Williston (1916b, p. 217) referred a skull in the Walker Museum to *Cardiocephalus* and presented the following distinctions:

1. *Gymnarthrus* has posterior teeth largest, *Cardiocephalus* those near the middle of the maxillary.
2. *Gymnarthrus* orbits relatively small, those of *Cardiocephalus* nearly equal to the distance from orbit to nares in diameter.
3. *Gymnarthrus* with large parietal foramen, that of *Cardiocephalus* small.

As early as 1910 Broom suggested that squamosal and quadratojugal were missing from the skull as a result of breakage. Close scrutiny of the *Gymnarthrus* skulls reveals that the supposed opening in the temporal region is the result of crushing, confirming Broom's opinion. On the left side of the paratype (A.M.N.H. 4763a, fig. 5E) broken edges of bone can be seen behind the orbit, along the outer edge of the postorbital and anterior edge of the supratemporal. Farther back the apparently "finished" edge of the left supratemporal coincides with the squamosal suture. A piece of the squamosal is present, crushed inward but projecting from the cranial roof above the exposed coronoid process of the jaw. On the right side the roof is more extensively broken, but the whole edge of the opening is clearly a fracture.

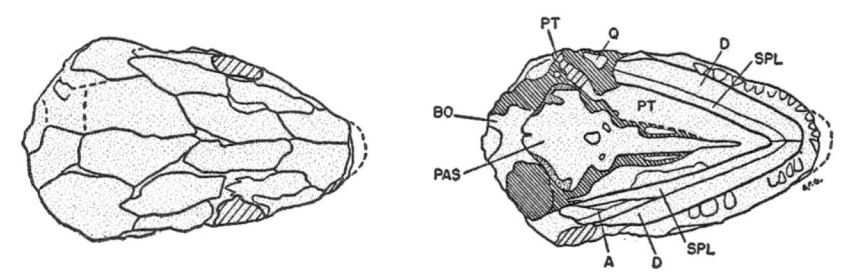

Figure 4. *Cardiocephalus sternbergi* Broili. Type skull of *Gymnarthrus willoughbyi* Case, A.M.N.H. no. 4892. Dorsal and palatal views. x 3. A angular, BO basioccipital, D dentary, PT pterygoid, Q quadrate, SPL splenial.

The type skull, A.M.N.H. no. 4892 (fig. 4), has the left postorbital better preserved than the paratype, but still broken off posteriorly. Remaining boundaries of the left skull "opening" are the lateral sutures of postfrontal and supratemporal. A small fragment of squamosal again is preserved.

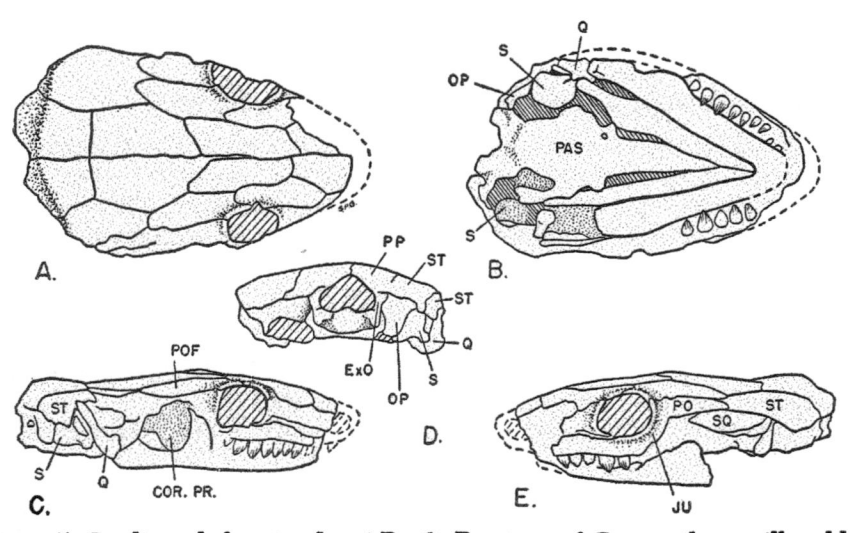

Figure 5. *Cardiocephalus sternbergi* Broili. Paratype of *Gymnarthrus willoughbyi* Case, A.M.N.H. no. 4763a. A. dorsal, B. palatal, C. right lateral, D. occipital, E. left lateral views. x 3. COR–PR coronoid process, EXO exoccipital, Q quadrate, S stapes, ST supratemporal, POF postfrontal, JU jugal, PO postorbital, SQ squamosal, OP opisthotic, PP postparietal, PAS parasphenoid.

On the right side the postorbital is preserved; it is similar in form to those of *Euryodus* and *Cardiocephalus*. Remnants of the squamosal are preserved here. All details of the skull roof pattern, the raised orbital rim, and the form of the teeth agree with the well preserved *Cardiocephalus* skulls from Oklahoma—and with Broili's figures of the type of that genus.

When the false temporal opening is eliminated, the resemblances between *Gymnarthrus willoughbyi* and *Cardiocephalus sternbergi* are so great that the possibility that these species are synonymous must be considered. Supposed distinctions between them, mentioned above, may be critically examined.

1. The excavation of the temporal region in *Gymnarthrus* is the result of crushing; in reality both genera had completely covered cheeks.
2. The parietal foramen is small and variable in size in these specimens; its apparent absence from Broili's type cannot be regarded as of systematic importance.
3. The *Gymnarthrus* types have orbital rims set off from the skull roof by grooves. These grooves are clearly what Broili described as "lyra." Case correctly stated that lyra—in the sense of entrenched slime canals—were absent from *Gymnarthrus*. The actual structures are the same and the supposed distinction a matter of mistaken terminology.
4. There are no cutting edges on the teeth of *Gymnarthrus* nor on those of *Cardiocephalus* specimens from Oklahoma. However the enlarged cheek teeth are compressed. Specimens from which matrix has been removed only to the middle of the teeth, as is commonly done on these tiny skulls, have the appearance of a sharp edge where none exists. I believe that such an illusion may have led Broili to report the keeled teeth. With so many other features in common, it is doubtful that his specimens were different in this respect.
5. The relative size of the teeth is somewhat variable as is shown by the detailed analysis of specimens from Oklahoma given elsewhere in this article. All agree in having large teeth beneath the orbit, and lack the abruptly enlarged and rounded tooth of *Euryodus*. The bulbous crowns of the *Gymnarthrus* teeth show the same longitudinal striations found on *Cardiocephalus* teeth. Broili did not report these on the type nor does Camp mention them; possibly they were eroded from his specimens, as they have been from several of the teeth of the *Gymnarthrus* and *Pariotichus* types.
6. Relative size of orbits. Our measurements of these specimens do not reveal significant differences in this character.

Accordingly we conclude that but a single species of a small gymnarthrid occurs in the Arroyo formation of Texas, and that its proper name is *Cardiocephalus sternbergi* Broili. Its distinction from the larger *Euryodus primus* Olson is discussed under that species.

A crushed and incomplete skull from the Arroyo formation, Walker Museum no. 1047, was figured by Williston (1916b, p. 219, fig. 63). It shows the right orbit and maxillary. Several vertebrae are crushed into the basicranial region. The nares are not visible. Lower jaws are present but their teeth are concealed medial to the upper dentition. The right maxillary bears 12 teeth, of fairly uniform height of crown, their diameter gradually increasing to the seventh tooth. The sixth and seventh teeth are very large, the eighth and ninth progressively smaller, and the tenth to twelfth are tiny teeth abruptly smaller than the ninth. Teeth appear laterally compressed, but not sharp edged. Radial striations are present on the crown of the sixth tooth.

The orbit has a well defined out-turned margin and circumorbital groove.

Professor Romer informs us that there is no positive evidence of association of the vertebrae with the skull. Thus Williston's statement (1916b, p. 217) that the neural spines of the cervical vertebrae are higher than the centra may not apply to *Cardiocephalus* and does not necessarily conflict with the evidence of isolated vertebrae from Oklahoma that neural spines were extremely low throughout the column.

CARDIOCEPHALUS CF. *STERNBERGI* BROILI FROM OKLAHOMA

Microsaur skulls from the fissure deposit north of Fort Sill, Oklahoma (Y.P.M. no. 3689, K.U.M.N.H. no. 8967, and the skull of O.U. no. 1034), are slightly larger than the cotypes of *Cardiocephalus sternbergi* and the type and paratype *Gymnarthrus willoughbyi*, but are similar in all other important features to these specimens. Although our present knowledge of these animals is far too meagre to establish adequate specific criteria, it may be said that there is no basis for regarding the Texas and Oklahoma forms as distinct. Their age is about the same, so far as may be judged by comparing *Captorhinus* from the Fort Sill deposit with species in the Arroyo formation (from which all other *Cardiocephalus* specimens have been obtained). The localities are not far apart and lay within the same land mass in early Permian time. Minor size differences alone are insufficient basis for specific separation.

TABLE 1

Comparative skull measurements in millimeters

	Length	Width
Cardiocephalus sternbergi Broili type [from figs.]	16.3	13
Gymnarthrus willoughbyi Case Type A.M.N.H. 4892	16±	10+
Gymnarthrus Paratype A.M.N.H. 4763a	16±	10±
Y.P.M. 3689	20	14.3
O.U. 1034	20−	16
K.U.M.N.H. 8967	20±	17±
Euryodus primus type	33	23

Y.P.M. NO. 3689, SKULL AND JAWS

The skull is 19.9 mm. long, 14.3 mm. wide in the temporal region, and 7.1 mm. deep, including the jaws, which are in place. The proportions are similar to *Euryodus*, an animal about twice its size, except that it is less swollen in the temporal region, the maximum width being reached farther back near the ends of the occipital crest, instead of near the junction of the postfrontal and supratemporal bones. The dermal bone pattern of the skull roof (fig. 6) is typically microsaurian (Romer, 1950, p. 634-635) with a

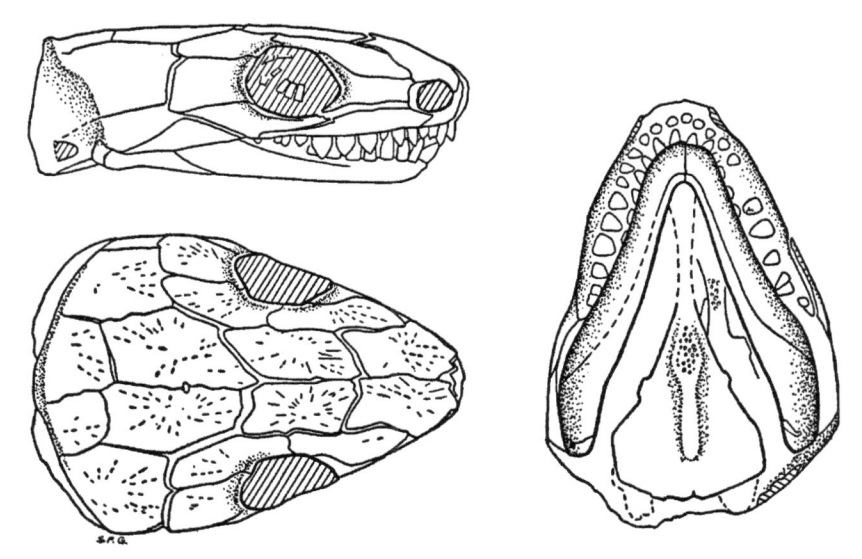

Figure 6. *Cardiocephalus* cf. *sternbergi* Broili. Y.P.M. no. 3689. Lateral, dorsal, and palatal views of skull. x 3.

large supratemporal reaching the rear margin of the skull roof, and there is no tabular; postparietals are well developed. Between the parietals is a minute pineal foramen. The large orbits lie just forward of the center of the skull. They are directed outward and slightly forward, and have slightly raised rims separated from the rest of the skull roof by shallow grooves. A septomaxillary forms the lower and posterior rim of each external naris and sends a slender process toward the center of those openings from its upper end. It extends backward beneath the anterior edge of the lacrimal.

The roofing bones are smooth save for sparse vascular pits which show a definite radial arrangement with respect to the centers of ossification.

The occipital surface of the skull meets the roof at a sharp angle, and slopes downward and backward to its lower limit above the foramen magnum and fenestrae ovales. It is convex from side to side, giving the rear margin of the skull roof a curved outline.

The braincase reveals little but can be interpreted on the basis of the types of *Gymnarthrus*. In A.M.N.H. no. 4763a the stapes is seen to form the posteroexternal corner of the braincase; large cavities in this region on the skull from Fort Sill are therefore the fenestrae ovales. The posterior end of the parasphenoid is forked, a projection extending beneath each of the paired condyles.

A small portion of the sphenethmoid is exposed in the space between parasphenoid rostrum and pterygoid, but the preservation of the specimen does not permit developing this element.

On the palatal surface (fig. 6), which was badly incrusted with pyrite and could not be completely prepared, the principal feature visible is the large parasphenoid which tapers rather gradually forward from beneath the otic region to a slender cultriform process bearing conspicuous basipterygoid processes near its base. A patch of small teeth occurs between the basipterygoid processes and may have extended forward along the cultriform process. A small portion of the palatal surface of the pterygoids has been exposed demonstrating that the interpterygoid vacuities were narrow. A fine shagreen of tiny teeth covered the palatal surface of the pterygoid. Details of the palatal dentition are shown by K.U.M.N.H. no. 8967 described below. The palatal structure is essentially that of *Euryodus* and other microsaurs, i.e., very conservative.

The jaws are shorter than the skull, as is usual in vertebrates with a forwardly sloping suspensorium. The midline distance from the front of the dental symphysis to the rear of the jaws is 15 mm. in comparison to the skull length of 20 mm. The jaws from tip of dentary to angle are 16 mm. long.

The articular region is low; as one may clearly see on the *"Gymnarthrus"* types, the upper border slopes downward from the high coronoid process to the articular cotylus. Behind the jaw articulation, the angular terminates in a rounded nob, in contrast to a distinct retroarticular process shown by University of Kansas specimens (fig. 8D).

The dentary is massive. A large angular extends from the rear of the dentary to the end of the jaw, and forms the ventral border of the posterior region. Its anterior end lies between the dentary and postsplenial.

Details of jaw structure, revealed by isolated specimens, are described farther on.

DENTITION: Each premaxillary bears three vertical incisors of equal size, round, blunt pointed, not enlarged. Nine teeth can be determined with certainty on the maxillaries, but 10 may well be present, especially on the right side. They increase rapidly in diameter from the first to the fifth (left) or sixth (right) which is a heavy, compressed, bulbous, blunt conical tooth with slightly trenchant anterior and posterior edges. The teeth are abruptly lower and smaller behind this, diminishing in size to the end of the series.

In the lower jaw the two most anterior teeth of each side are somewhat procumbent; the third is similar to them in size. These incisors are slender cylindrical teeth, with abruptly blunt conical tips; presumably the slightly worn upper incisors were of similar form. Beginning with the fourth lower tooth the diameter increases, and the exposed portion suggests a shape similar to the mid maxillary series back to the seventh, which lies just in front of the largest of the upper series. There is some indication that the eighth may have been the largest of the lower series; behind that they could not be exposed because of the overlapping jugal bones.

Additional data on the lower dentition is supplied by several lower jaws found free among the bones in the soft clay. They reveal marked variability in number of teeth and especially in the position of the largest tooth in the jaw and number of enlarged "crushing" teeth. Y.P.M. nos. 3681 (fig. 7) and

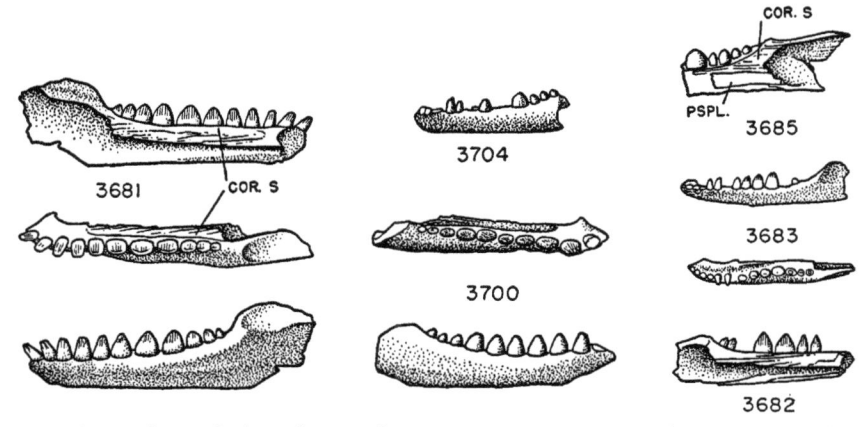

Figure 7. *Cardiocephalus* cf. *sternbergi* Broili. Lower jaws, Y.P.M. no. 3681 internal, occlusal, and external views; Y.P.M. nos. 3685, internal view; 3683, external and occlusal views; 3704, external view; 3700 occlusal and external views; 3682, internal view. All x 3. COR–S coronoid suture, PSPL postsplenial.

3702 have 11 teeth, of which the eighth is the largest, the sixth to eighth all enlarged, bulbous, and laterally compressed. All teeth from the first incisor through the eighth tooth are of even height, posteriorly they are abruptly lower, though the ninth is still somewhat bulbous. A small dentary, Y.P.M. no. 3704 (fig. 7), has 12 teeth preserved; it is so broken at the rear that it cannot be determined whether all teeth are present. The eighth and ninth teeth are much enlarged and compressed, the three (or more) behind the ninth low and small. In front of the eighth the teeth decrease in diameter, those in front of the sixth being slender and irregular. Y.P.M. no. 3683 (fig. 7) has 14 teeth, of which the eleventh is largest and the tenth also considerably enlarged in comparison to the ninth. As in 3681 the height of crown is quite even regardless of diameter in the anterior teeth. It is worthy of note that these two jaws (nos. 3683 and 3704) differ from the others in smaller size as well as more numerous teeth. Two fragments, Y.P.M. nos. 3682 and 3685 show respectively three and four small compressed shearing teeth behind the large cheek tooth. Still another jaw, no. 3700, with 10 teeth, has all the anterior teeth equally enlarged and bulbous, the first procumbent, the second slightly longer than the rest as though caniniform; thence a uniform series to the seventh, behind which the last three teeth are abruptly smaller. In all specimens the crowns of all teeth show delicate striations radiating from the apex, and, where present, recurved tips on the slender anterior teeth. The first incisor is procumbent, the second somewhat so.

Tooth counts made on a total of 24 dentaries and maxillaries of *Cardiocephalus* in the Kansas, Yale, and Chicago collections from Fort Sill are given below. It is evident that the count is variable but usually within relatively narrow limits. Although the maximum number of mandibular teeth exceeds that of *Euryodus*, there is overlap in numbers and the genera must be separated by morphology of the dentition rather than tooth counts.

TABLE 2

Tooth counts of *Cardiocephalus*

Upper teeth (premaxillary and maxillary)

	Total	Largest tooth
K.U.M.N.H.	11	8
	11 (12)	9
	11 (12)	8 (9)
	11 (12)	8 (9)
	12	8
	12	9
	12	9
	12	9 (10)
Y.P.M.	12	9

Lower teeth

	Total	Largest tooth
K.U.M.N.H.	10	7
	11	7
	12	8
	12	8 (9)
	12	9 (10)
	?	9
	?	9
C.N.H.M.	10	8
	10	8
	?	9
	?	9 (10)
	13	10
Y.P.M.	11	8
	13	11
	15	11

K.U.M.N.H. no. 8967, skull with lower jaws, figure 8, A-D

The specimen is slightly crushed and damaged by weathering of the posterior end. It was encased partly by crystals of transparent calcite and partly by soft clay. There is slight damage by pyritization which obscures the anterior palate, and hematite lines the sphenethmoid cavity ventrally. Otherwise the detail remaining is excellent.

In dorsal view the snout and orbital regions are complete back to the parietals, postfrontals, postorbitals and in part the jugals. A parietal foramen is barely indicated by a broken corner of the right parietal and by impression in the underlying calcite matrix. In ventral view, the lower jaws are slightly separated at the symphysis but otherwise nearly in position; the right ramus is virtually complete, showing with exceptional clarity the dentary, splenial, and postsplenial, angular, surangular, articular, and prearticular, with three coronoids also present but hidden from view. Hidden details are supplied by an isolated left ramus (K.U.M.N.H. no. 9923, fig. 8, E-G).

The palate preserves nearly perfectly the complete right pterygoid and most of the left including a central portion of the parasphenoid-basisphenoid complex showing the right basipterygoid process perfectly preserved and slightly disengaged from its socket in the pterygoid. The latter is strongly articulated with its quadrate. The anterior extent of the parasphenoid is broken but it clearly underlies a well ossified and extensive sphenethmoid. In the floor of the right orbit may be seen a dorsal view of the ectopterygoid, the posterior end of the palatine, and part of the lateral border of the pterygoid.

The marginal dentition is completely preserved except where it is broken off the left premaxillary. Careful excavation through the floor of the right orbit exposed a few coronoid teeth otherwise hidden from view but undoubtedly like those of the isolated left ramus (fig. 8, F, G). Parasphenoid denticles occur between the basipterygoid processes and numerous pterygoid denticles, sharply conical and inclined posteriorly, occur back to the level of the basicranial articulation.

The occiput, otic capsules, and posterior skull table are missing. As in the type of *"Gymnarthrus,"* the cheek region of the K.U. skull appears to be excavated thus producing in lieu of other evidence (which now exists) the erroneous impression of a "naked" articular region.

The roof of the K.U. skull, insofar as it is preserved, corroborates perfectly the characteristics of the Yale skull and in addition shows more clearly the nature of the lacrimal and septomaxillary. The latter is a spirally curved bone which forms the posterior and ventral rim of the external naris and extends medially in complicated fashion. The lacrimal, shown by both the skull and a perfect, isolated specimen (K.U.M.N.H. no. 9925), has a nasolacrimal canal originating from two or three foramina in the anteroventral orbital wall close to the orbital rim and running as in *Captorhinus* through the thickened lateroventral base of the bone to the posteroventral

20 REVISION OF THE GYMNARTHRIDAE

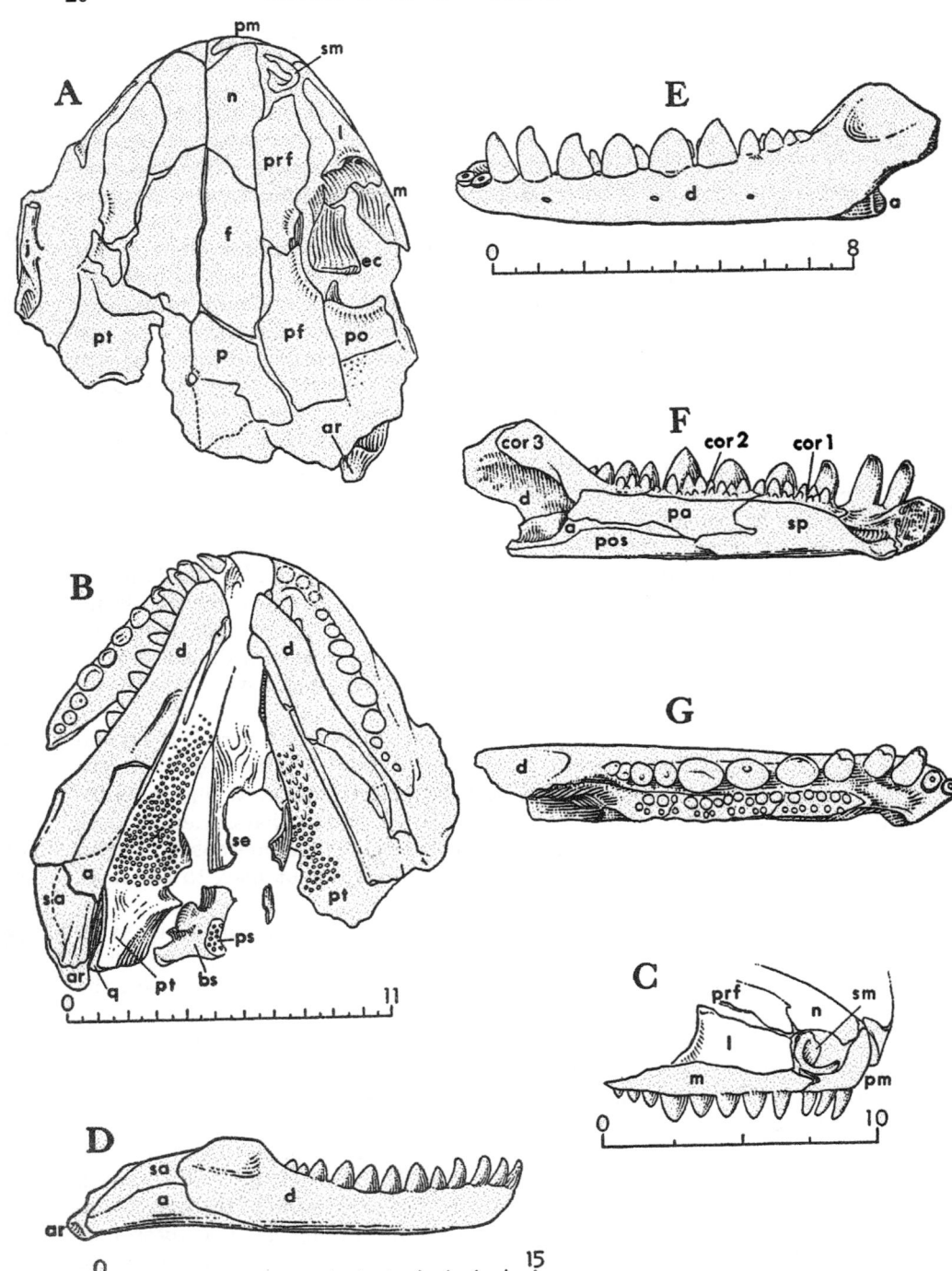

rim of the naris. Here the duct may have opened through the body of the septomaxillary. The lacrimal has an extremely broad base posteriorly, possibly correlated with the wide, low skull and powerful dentition; the ventral surface extends far medially to overlap dorsally a relatively large area of palatal bone. The premaxillary, as shown by two isolated specimens, is the characteristic three-pronged structure but differs from associated *Captorhinus* in its more delicate structure, and in having an unforked nasal process which with its paired fellow wedges between the nasals. Also there are three teeth rather than four or more as in *Captorhinus*.

The pterygoid extends far anteriorly on the palate to at least the level of the third maxillary tooth. The median border of the palatal process is gently concave thus forming a modest interpterygoid vacuity. The pterygoid is broader medially than it appears to be in the Yale skull. Posteriorly the pterygoid clasps the quadrate strongly as in *Pantylus* (Williston, 1925, fig. 7). The basisphenoid articulation in the pterygoid is a deep notch set obliquely to engage the basipterygoid process strongly. Presence of a bony epipterygoid involved in the articulation can not be recognized. Laterally and anterior to the level of the articulation there appears to be a distinct suture with an elongate, somewhat downturned bone forming the anterior half of the median border of the subtemporal fossa. This seemingly separate bone is too far posterior to be an ectopterygoid and is probably only a broken part of the pterygoid where it normally forms the median border of the subtemporal fossa. The palatal bone exposed in the floor of the orbit appears to be a large ectopterygoid in normal relationship with the posteromedian border of the maxillary (here slightly separated). Anteromedially in the orbital floor another bone, probably the palatine, is exposed in a position which suggests what the ventral surface of the pterygoid indicates —namely that the palatine does not have an excessive palatal expanse as shown in *Pantylus* (Williston, 1925, fig. 6). The condition in the latter may be misrepresented.

Details of the palatine and vomer are completely hidden from below. Possibly these bones present a toothed surface to the elongate battery of teeth on the coronoids of the lower jaw. This detail cannot be determined clearly, but there appears to be at least two large teeth on the ectopterygoid.

The general pattern of the dermal bones of the palate appears to be that of primitive labyrinthodonts. There is no suggestion of a transverse flange developed on the pterygoid as in the little stem reptile, *Captorhinus*.

Figure 8. *Cardiocephalus* cf. *sternbergi* Broili. A–B., dorsal and ventral views of skull and jaws, K.U.M.N.H. no. 8967. C. oblique view of external naris of same. D. lower jaw belonging to this skull with missing parts restored. E–G., lateral, medial, and dorsal views of isolated lower jaw, K.U.M.N.H. no. 9923. Millimeter scale in drawing. a angular, ar articular, bs basisphenoid, cor 1, cor 2, cor 3, coronoids, d dentary, ec ectopterygoid, f frontal, j jugal, l lacrimal, m maxillary, n nasal, p parietal, pa prearticular, pf postfrontal, pm premaxillary, po postorbital, pos postsplenial, prf prefrontal, ps parasphenoid, pt pterygoid, q quadrate, sa surangular, se sphenethmoid, sm septomaxillary, sp splenial. Drawn by Frank E. Peabody.

Enough of the ossified neurocranium remains in the K.U. skull to delineate the contours fairly completely. A nearly complete, isolated basicranium of *Euryodus* (fig. 16, H, I) fills in needed detail. The basisphenoid presents to the pterygoid a strong basipterygoid process with a finished, hemicylindrical, articular surface inclined obliquely from above, downward, and posterolaterally. The parasphenoid is not clearly separate, but the usual patch of denticles indicates the extent of this bone between and behind the level of the basipterygoid processes. On the part of the basisphenoid preserved, there is a small foramen between parasphenoid and basisphenoid for entrance of the internal carotid artery. The canal courses medioanteriorly and enters the lateroventral aspect of the pituitary pit. The latter is exceptionally deep, and at its rear end it is set off from a transverse groove which carried a large interorbital vein. The isolated basicranium from *Euryodus* (fig. 16) agrees in all particulars with that of the K.U. *Cardiocephalus* skull except that it is larger (6 mm. wide across the basipterygoid processes compared to 3.5 mm.). In lateral view the basisphenoid shows a sturdy, lateral wall extending forward to meet an equally sturdy sphenethmoid. The wall is pierced just above the position of the basipterygoid process by a large, ovoid foramen for the interorbital vein. The groove across the back of the pituitary pit, noted above, is roughly the same calibre as the ovoid foramen. The wall flares lateroposteriorly to meet the prootic—here missing. A canal for the abducens nerve enters the basal plate of the basisphenoid and courses forward to an exit just posterior to the large opening for the interorbital vein.[*]

Anteriorly the parasphenoid and sphenethmoid appear indistinguishably fused. The lateral wall of the sphenethmoid, clearly exposed by damage to the skull roof (fig. 8, B), was exceptionally well ossified and powerfully built in view of the smallness of the skull—probably correlated with the powerful jaws—and extended upward nearly if not to the skull roof. The maximum thickness of the wall is approximately 0.25 mm. On both sides of the midline the walls are incomplete and slightly displaced in a narrow, transverse zone separating basisphenoid and sphenethmoid regions, but whether the presence here of foramina or of cartilage caused the break, one cannot be certain. The sphenethmoid region is broader than indicated in the Yale skull, but has the proportions of *"Gymnarthrus"* (A.M.N.H. no. 4763). The breadth of this region suggests the broad platybasic plan of the urodele basicranium.

LOWER JAW: The lower jaw of *Cardiocephalus* (fig. 8, D–G) is powerfully constructed; a prominent coronoid process is developed on the dentary,

---

[*] Six isolated specimens of a basisphenoid—parasphenoid complex in K.U. collections are comparable to *Cardiocephalus* in size but are clearly different, for example, the cultriform process is narrow, the parasphenoid denticles form a triangular platform, the basipterygoid process is not fully ossified, and there is no definitive pit for the pituitary. The six specimens may represent a different group of small amphibians of the same size as *Cardiocephalus*, but the large number of specimens of this rare element in the collection is puzzling.

and impressed on its laterodorsal surface is a shallow "masseteric" fossa. By contrast, *Euryodus* has a much deeper fossa which faces directly laterally, and the coronoid process is higher and longer. The position of the articular surface is relatively low with respect to the coronoid crest and tooth row thus increasing the leverage of adductor muscles of the jaw. There is a prominent retroarticular process—1 mm. long compared to 15 mm. total length of the jaw ramus. Bone surfaces are smooth; no sculpturing or grooving is evident.

The dentary in lateral view appears to dominate the jaw structure and has a superficially therapsid-like appearance especially in specimens of the related genus, *Euryodus*, where the "masseteric" fossa is deeply excavated and the coronoid process of the dentary sweeps far back. (The swollen temporal region of *Euryodus* is probably correlated with a more powerful masseter-like division of the adductor muscles.) The dentary extends to the ventral border of the jaw thus excluding the splenials from the lateral surface as in advanced reptiles and urodeles.

Some dentaries of *Cardiocephalus* are massive and deep, considerably stouter than the jaws of *Euryodus*. However in the smallest jaws, Y.P.M. nos. 3683 and 3704 (fig. 7), the dentary is slender and delicate. Posteriorly a gently arcuate angular is overlapped by the dentary and in turn overlaps an equally arcuate surangular. A relatively large, lateral area of the articular is left exposed at the posterior ends of surangular and angular.

In median view a splenial, engaging in the ventroposterior area of the symphysis, narrows posteriorly to a slender neck and then expands abruptly to a broad, bifurcate blade which buttresses the anterior coronoid platform. Judging from sutural surfaces on well preserved dentaries, the splenial of *Euryodus* does not enter the symphysis to the degree seen in *Cardiocephalus*. Also the symphysial surface of the dentary of *Euryodus* differs by having a deep Meckelian groove. The slender neck of the splenial of *Cardiocephalus* fails to cover the anteriormost portion of the Meckelian canal. The dorsal and ventral posterior processes of the splenial blade overlap a long prearticular above and a splint-like postplenial below. The dorsal process and the anterior half of the prearticular buttress the lingual border of the coronoid platform as in *Pantylus* (Williston, 1925, fig. 18) so that the coronoid dentition is a powerful accessory to the marginal dentition. There are three coronoid bones, each bearing teeth and separated by oblique sutures running from the lingual border, anterolaterally. The posterior coronoid curves upward along the median surface of the coronoid process of the dentary and forms the anterior border of the adductor fossa. Thus the coronoid dentition of *Cardiocephalus* is powerfully developed and is indicated on the many isolated dentary bones by the characteristic, long coronoid suture reaching far forward on the dentary medial to the marginal teeth. The related *Euryodus* has no comparable development of this suture on the dentary and must be presumed to have had much less development of at least the anterior coronoid dentition. The prearticular is narrow and long—extending over two-thirds the length of the ramus. Below the articular region, the prearticular flares medially forming an elongate, ventromedian depression. The angular forms the ventral border of the posterior third of the ramus and extends forward under the postsplenial in a long squamous suture.

No large foramina or fenestrae other than the adductor fossa are visible on lateral or median surfaces of the ramus. Small nutrient foramina occur sparingly along the lateral surface of the dentary; in the larger *Euryodus* these are larger and more numerous.

The structure of the lower jaw is fundamentally that of an amphibian, but with reptilian features, and resembles closely the controversial *Pantylus*. The major differences between the lower jaw of the latter and that of *Cardiocephalus* (and *Euryodus*) lies in specialization of the dentition, particularly the transfer of major function from marginal to the coronoid dentition.

O.U. NO. 1034, ARTICULATED SKELETON

Through the kindness of the late Professor J. Willis Stovall of the University of Oklahoma, and by courtesy of Professor Alfred S. Romer of Harvard University (who also has studied this specimen but most kindly turned it over to us for description) it has been possible to examine and further expose this rare skeleton. Owing to the friable nature of the bone and rather brittle calcite matrix, full preparation was not attempted, but several characteristic regions of the vertebral column have been exposed.

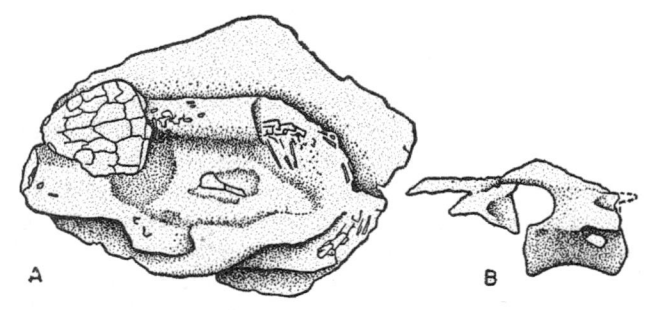

Figure 9. *Cardiocephalus* cf. *sternbergi* Broili. A. nodule containing articulated skeleton, O.U. no. 1034. x 1. B. lateral view of mid-dorsal vertebrae of the same specimen. x 2 1/2

SKULL: The skull (fig. 9A) has been somewhat flattened posteriorly so it appears broader in the temporal region than other specimens. Limits of cranial bones are clearly marked by matrix-filled sutures. The orbits have weakly outturned rims and conspicuous circumorbital grooves except on the postfrontal. A transverse crack strongly resembling a suture parallels the occipital crest about one-fourth millimeter in front of it across the entire skull roof. That this cannot be a suture is demonstrated by its absence from other *Cardiocephalus* skulls with otherwise similar roof patterns, and by the impossibility of homologizing bones behind a suture in this position.

The occipital crest is slightly raised. Occiput and skull roof form an abrupt obtuse angle. Sutures between postparietals and supratemporals continue downward on the occipital surface as far as it is exposed.

VERTEBRAE: The atlas is represented by two short, tall, stout, rod-like neural arches which converge upward. Probably they originally were united dorsally as indicated by K.U.M.N.H. specimens; the incomplete ends emerge from the concretion separately. It has not been possible to expose their articulation with the skull but this detail is provided by isolated K.U.M.N.H. specimens; a slight gap separates their upper end from the foramen magnum and forward sloping occiput.

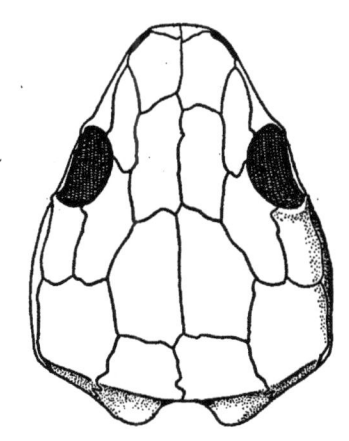

Figure 10. *Cardiocephalus* cf. *sternbergi* Broili. Restoration of skull of O.U. no. 1034, by L. I. Price. x 3.

The neural spine of the second vertebra is broad and rather short. It does not rise appreciably above the zygapophyses of the third vertebra. Anteriorly traces of prezygapophyses may be distinguished and just lateral to them a transverse process, high on the side of the neural arch. So far as may be determined, it was unspecialized, similar to other dorsal vertebrae, like the second vertebra of urodeles or caecilians. The first rib is two headed and slightly stouter than those which follow; its shaft curves back parallel to succeeding ribs.

Two groups of dorsal vertebrae were partially exposed, as shown on figure 9A. Both reveal low, broad, flat-topped neural arches (fig. 9B), as wide as the zygapophyses throughout their length, and bearing a minute rounded median ridge in place of a neural spine. They differ from *Captorhinus* vertebrae in the absence of the inverted V-shaped attachment area for the interspinous muscle or ligament, and lack the swollen postzygapophyses of cotylosaurs. Transverse processes are short and anteriorly placed. The centra are concave laterally and swell to round, closely appressed articular faces, with no trace of intercentra. Ribs are two headed, the capitulum extending toward the intervertebral articulation, the tubercle toward the transverse process. Shafts of the ribs are extremely slender and somewhat curved.

Characteristics of vertebrae are but poorly revealed by the imperfect preparation of this specimen. Sufficient detail is shown, however, to permit certain identification of the isolated vertebrae washed from clay of the fissure fillings with this animal. These vertebrae are described on a later page.

LIMB BONES: Traces of the left humerus and proximal end of the ulna are preserved. The humerus has a primitive 90° twist between its expanded ends but a relatively long and slender shaft. The form of the head and deltoid process, as well as its general size and proportions, are similar to the numerous free humeri recovered from the same deposit, and render their identification as *Cardiocephalus*-sure.

It has not been possible to expose the caudal vertebrae or rear limb of this specimen.

EURYODUS PRIMUS OLSON

TYPE: W.M. no. 1565 and paratypes W.M. nos. 1566 and 1567 skulls.
LOCALITY: Brushy Creek, Baylor County, Texas.
(Referred skeletal elements from Fort Sill locality.)
AGE: Arroyo formation (Clear Fork Group).

A larger gymnarthrid from the Clear Fork beds of Texas has been described by Olson (1939, p. 389–397) and additional disarticulated material is known from the fissure deposits of Fort Sill. Its general appearance is so similar to *Cardiocephalus* that at first the possibility that it merely represented larger, more mature individuals of that genus was considered. There are, however, a number of distinctions which indicate that it is a separate and valid genus.

Euryodus differs from *Cardiocephalus* in the following ways:

1. Larger size (skull length of type 33 mm. compared to 16–20 mm. in *Cardiocephalus*).
2. Coronoid short, probably toothless, narrowing anteriorly and rolling under the lingual border of the dentary (fig. 12). In *Cardiocephalus* the coronoids are toothed, and extend far forward against the lingual border of the dentary.
3. Cheek teeth, including greatly enlarged cheek tooth are round, not (or very little) laterally compressed. Olson (1939) represents the teeth as laterally compressed but the type has rounded teeth.
4. Tips of anterior teeth not recurved.
5. Height of crown increases rapidly from first tooth to large posterior tooth. (In *Cardiocephalus* teeth vary in diameter but are nearly the same height except the small posterior teeth.)
6. Jugal relatively large, rising higher behind the orbit and forming more of the postorbital covering of the cheek. The maxillary and jugal are flexed outward and upward from the tooth row, related to the swollen

cheek region, whereas in *Cardiocephalus* the lateral border is straight in ventral views.
7. A more pronounced masseteric fossa is developed in the coronoid process of the dentary.
8. Dorsal surface of skull more strongly pitted than in *Cardiocephalus*.
9. Orbital margin not protruding as rim.

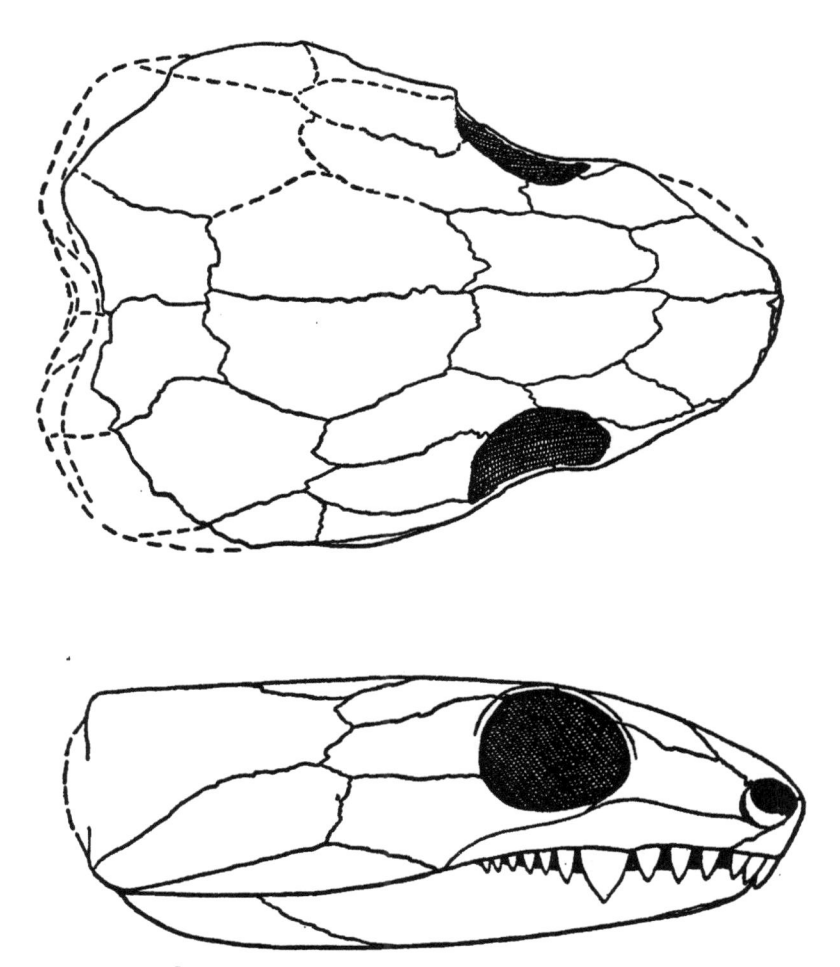

Figure 11. *Euryodus primus* Olson. Type skull, W.M. no. 1565. A. dorsal surface with outline restored. x 3. B. restoration lateral surface of skull by L. I. Price. x 3.

Of these features, the seventh is quite possibly related to the larger size of the animal. The outward swelling of the cheek and deep "masseteric fossa" in *Euryodus* are correlated with stronger jaw musculature associated with the pronounced enlargement of a single specialized crushing tooth. Other distinctions in dentition, jaw structure, and size, seem quite constant and therefore significant of specific and generic distinction.

Collections of the University of Kansas contain numerous isolated bones of *Euryodus,* which is associated with *Cardiocephalus* at Fort Sill. The two genera appear to be closely similar postcranially and also in the basicranial and occipital regions. The only reliable criteria in these particular regions is the difference in size. A series of more than 50 dentaries and 18 maxillaries illustrates the marginal dentition of the larger form. Unexplained is the dearth of other elements assignable to *Euryodus.* On the basis of frequency of occurrence of dentaries and maxillaries, the two genera are about equally abundant at Fort Sill. But other elements of *Euryodus* seem to be much rarer, and not as a result of confusion of these larger elements with smaller individuals of the common captorhinid species since the latter have been studied closely.

Certain notes on the cranial anatomy of *Euryodus* supplementary to Olson's description may be offered here. Postparietals are paired elements (fig. 11A), not single as shown in Olson's original drawing (1939, p. 391, fig. 1). A septomaxillary is present within the narial opening; the teeth are rounded rather than laterally compressed.

Olson mentions a "very small otic notch." No structure which we would so identify can be found; possibly the slight concavity of the posterior border of the temporal region was so interpreted. This is doubtful, however, for the structure of the gymnarthrid middle ear suggests the tympanic membrane had been lost as in urodeles or caecilians.

In the basicranial region the basipterygoid processes lie farther forward than suggested by Olson's figure, and the parasphenoid widens behind them, underlying the braincase, to three times its anterior width. In general, this region closely resembles that of *Cardiocephalus.*

Occipital flanges of postparietals and supratemporals cover the upper surface of the rounded occiput.

The mandibular dentition of Walker Museum no. 1567 shows a large tooth corresponding in position to the enlarged maxillary tooth, followed by three small teeth. W.M. no. 1569 is the anterior portion of a mandible bearing seven or eight (?) teeth. These are of equal size (diameter?) and comparable to the three teeth preceding the enlarged tusk of the maxillary. Specimens from Oklahoma referred to this genus vary from nine teeth, of which the seventh is largest, to 13 teeth with the tenth greatly enlarged.

The accompanying table shows tooth counts on all sufficiently complete specimens available. It may be significant that *Euryodus primus* from Texas differs from the Fort Sill specimens in having a greater number of small teeth (6) behind the enlarged upper tooth. Perhaps the Texas and Oklahoma specimens are not conspecific.

Many vertebrae are present in the nodular masses accompanying the type skull of *Euryodus.* Price exposed and illustrated one of these (W.M. no. 1569, fig. 19) which is typical of all observed. The centrum is a single biconcave ossification, pierced for the notochord, its ventral surface smooth and broadly rounded. The neural arch is firmly attached to the centrum by a suture, traces of which are evident on cross sections of vertebrae. Some specimens show a slight swelling along the outer surface in the region of the neurocentral suture. Stout transverse processes arise entirely from the anterior portion of the neural arch, much like those of *Cardiocephalus*. As in that genus the zygapophyses extend well beyond the rim of the centrum. The neural spine is short, consisting of only a thin crest on the dorsal surface of the neural arch. Aside from size, these vertebrae are scarcely to be distinguished from those of *Cardiocephalus*. Possibly the large sacral vertebrae, mentioned on page 46, pertain to *Euryodus*.

The atlas, except for its greater size, is like that of *Cardiocephalus*, this conclusion being based on isolated specimens, K.U.M.N.H. no. 9926 and C.N.H.M. no. UR-287.

PANTYLUS, OSTODOLEPIS, AND *GONIOCARA*

It is not the purpose of this article to review the anatomy of *Pantylus*

TABLE 3

Tooth counts of *Euryodus* specimens

Upper teeth (premaxillary and maxillary)

	Total	Largest tooth
K.U.M.N.H.	10+	9
	12+	11
C.N.H.M. (type)	13	7

Lower teeth

	Total	Largest tooth
K.U.M.N.H.	10	8
	11	8
	11	8
	11	8 (9)
	11	9
	11	9
	11	9
	11	9
	?	8 (9)
	?	8 (9)
	?	9
C.N.H.M.	9	8
	11	8
	11	9
	11	9
	12	9
	12	9
Y.P.M.	9	7
	9 ?	7

REVISION OF THE GYMNARTHRIDAE

and its allies, or the varied opinions which have been expressed concerning their affinities. At the present time A. S. Romer and John A. Wilson are studying new and relatively complete material of *Pantylus* from the Putnam formation of Texas. A preliminary report (Wilson, 1951, p. 1490–91) suggested reptilian affinities, but Romer writes (March 6, 1955) that more

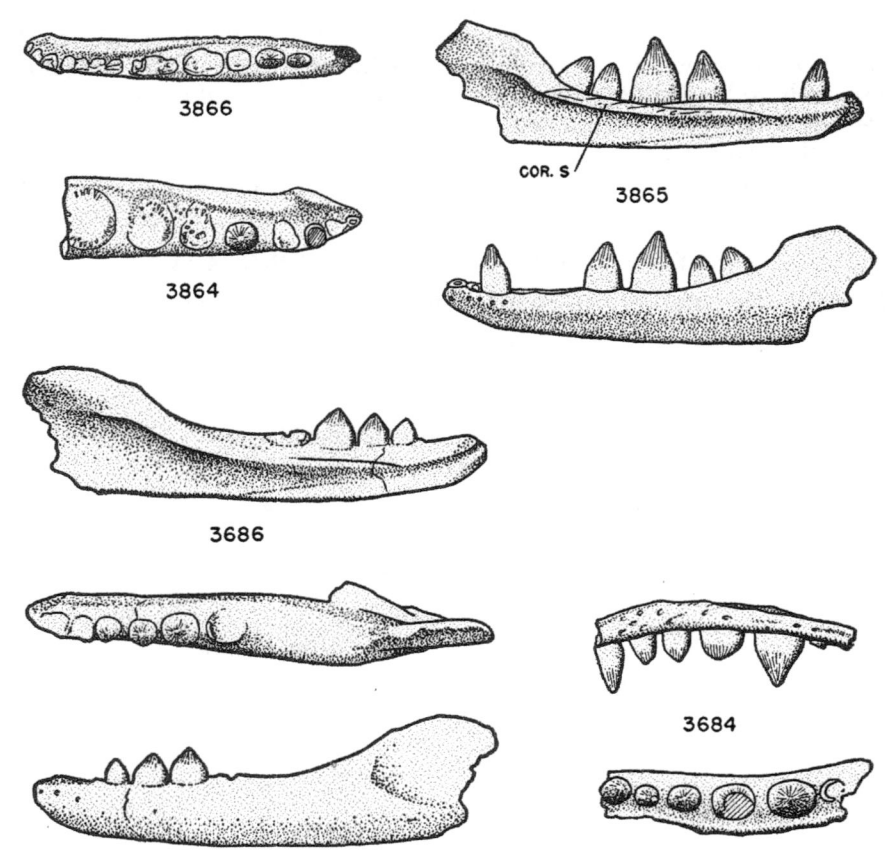

Figure 12. *Euryodus primus* Olson. Dentition. Lower jaw, Y.P.M. no. 3686, medial, occlusal, and lateral views. Maxillary, Y.P.M. no. 3684, lateral and occlusal views. Lower jaws, Y.P.M. nos. 3866, 3864, occlusal views; Y.P.M. no. 3865, medial and lateral views. Note anterior restriction of coronoid suture. x 3.

thorough preparation has shown agreement with the gymnarthrids, as here defined, in almost every detail.

PANTYLUS CORDATUS COPE (1881, p. 79)

The cotypes, two skulls, A.M.N.H. nos. 4330 and 4331, were collected on the Big Wichita River in Texas by Jacob Boll in 1880, from somewhere within the Wichita group. Mehl (1912) described another skull from the Big Wichita, and Williston (1916a, p. 165) discussed a group of specimens from Mitchell Creek, a locality within the Clyde formation in the upper part of the Wichita group. Romer (1935, p. 1613) refers to *Pantylus* as a new member of the Clyde fauna, but the more recently discovered specimens from the Putnam formation south of Scotland, Archer County, Texas, show that the genus was present earlier; they may well be conspecific with Cope's types.

Pantylus differs from both *Cardiocephalus* and *Euryodus* in its larger size, and particularly in the broad palatines and great development of crushing teeth on palate and coronoids. The marginal dentition is correspondingly reduced. The jugal is greatly expanded behind the orbit, more closely resembling captorhinid reptiles than either *Cardiocephalus* or *Euryodus*.

PANTYLUS COICODUS COPE (1896, P. 450)

This species was based upon a broken skull and associated teeth, A.M.N.H. nos. 4335 and 4336. They were obtained from the Arroyo formation (lower Clear Fork Group) on Coffee Creek in Baylor County, Texas. Case (1911a, p. 61) regarded it as doubtfully distinct from *P. cordatus*. It is not certain that it belongs to *Pantylus*.

PANTYLUS TRYPTICHUS CUMMINS (1908, P. 743, NOMEN NUDUM)

A small fragment of a lower jaw, A.M.N.H. no. 4445, from Coffee Creek, was so labeled by Cope; Case (1911a, p. 114) suggests that it belongs to an old individual of *Captorhinus*.

Thus the occurrence of *Pantylus* in the Clear Fork group has not been established, and this genus, the largest of the pariotichids, may well be confined to the Wichita beds.

OSTODOLEPIS BREVISPINATUS WILLISTON (1913, P. 363–366)

The type, Walker Museum no. 680, consists of a series of seven vertebrae and ribs, now missing. The type locality is West Coffee Creek, Wilbarger County, Texas, in the Arroyo formation, Clear Fork group. Case (1929) described and referred to this species as skeleton (U.M.M.P. no. 11156) which, "in all probability came from the same locality and geological horizon as the type specimen" (1929, p. 82). Its triangular skull has an unusually high postorbital region and a low, pointed snout, suggestive of an unusual adaptation, possibly burrowing. The maxillary teeth are more numerous and more slender and sharp pointed that those of *Pantylus* or the typical pariotichids.

Case regarded it as an aberrant cotylosaur, but also recognized amphibian characteristics.

The vertebrae resemble those of *Pantylus* so closely that Williston (1916a, p. 174) concluded that *Ostodolepis* was a synonym of the former. The peculiar skull of the animal described by Case is different from *Pantylus* and may well belong to *Ostodolepis*, especially if *Pantylus* does not occur in the Arroyo formation. Gymnarthrid features of this animal include the high coronoid process and large supratemporal.

GONIOCARA WILLISTONI (BROILI)

Two occipital fragments of small skulls from the Craddock Bonebed (Arroyo formation, Clear Fork group, L. Permian, 150 feet above Leuders limestone), Baylor County, Texas, were described by Broili (1913, p. 98–100, Pl. 9, figs. 3–4) as *Goniocephalus willistoni*. Hay (1929, p. 835) pointed out that the generic name was preoccupied [by *Goniocephalus* Kaup 1827] and substituted *Goniocara* for these fossils. Broili compared the form with placodonts and anomodonts, yet regarded it as an amphibian. He described the elements at the angle between occipital and dorsal surfaces of the skull as parietals because of an opening similar to a parietal foramen which is present between them in both specimens. This anomalous position for the parietals suggests that perhaps this opening is not truly a parietal foramen, and that the bones labeled parietals are really postparietals. The supratemporals appear broken away along their median sutures.

The genus is indeterminate from published information. The size, and abrupt angle at the back of the skull roof are slightly suggestive of *Pantylus* or *Ostodolepis*.

SYSTEMATIC SUMMARY

CLASS AMPHIBIA
SUBCLASS LEPOSPONDYLI
ORDER MICROSAURIA

FAMILY GYMNARTHRIDAE CASE 1910, P. 177.

Until recently it would have been necessary to abandon the generally used name Gymnarthridae (Case, 1911b, p. 14; Williston, 1916b, p. 218–219; Romer, 1945, p. 592; 1950, p. 638–639) for these microsaurs because *Gymnarthrus* proves to be a synonym of *Cardiocephalus*. Under these circumstances either a new family name based upon *Cardiocephalus* could have been proposed or the family Pariotichidae (Cope, 1883, p. 631) could have been redefined. The latter procedure might have led to confusion because Pariotichidae has been used largely, even by Cope himself and subsequently by Case (1911a, p. 33) and Williston (1916b, p. 219) for captorhinid reptiles. In 1953 the Fourteenth International Congress of Zoology at Copenhagen adopted revisions in the International Rules of Zoological Nomenclature permitting retention of a family name based upon a generic name which is a junior synonym of some other generic name [Copenhagen Decisions on Zoological Nomenclature, p. 36, par. 54 (1) (a)]. This change in the code makes it possible to retain the familiar Gymnarthridae as the name for this family of microsaurs.

DIAGNOSIS: Microsaurs with skulls less triangular and more elongate than the Microbrachiidae but less elongate than the Adelogyrinidae; orbits relatively large, just in front of middle of skull. Parietal foramen very small. Without supratemporal horn or otic notch. Postparietals well developed on skull roof; supratemporals larger than squamosal. Surface of skull bones smooth. Basioccipital small, not entering condyles. Teeth in single series on dentary, premaxillary, and maxillary; inflated, bulbous, round or compressed, with bluntly conical crowns often striated near tip. Numerous denticles present on coronoid(s) of some genera; small palatal teeth numerous on pterygoid and on parasphenoid between and posterior to basipterygoid articulation.

Vertebrae notochordal, centra round below, evenly constricted in middle, without intercentra; neural arches delicate, wide, flat topped, lacking neural spine; transverse processes of dorsal vertebrae near front of arch, short, widest at posterior edge; zygapophyses not inflated. Atlas of urodele type. Caudal vertebrae not certainly known. Scapulocoracoid a single ossification with screw-shaped glenoid, pierced by supraglenoid, coracoid, and glenoid foramina; dermal girdle unknown. Humerus slender, its ends expanded at right angles, without epicondylar foramina; radius and ulna weak. Pelvis with single puboischiadic plate; spine of ilium slender, curving posteriorly. Femur straight, with prominent inner trochanter, adductor ridge, no fourth trochanter. Tibia and fibula well developed. Feet unknown. Late Pennsylvanian and early Permian, North America.

KEY TO GENERA:

I. Small or medium sized. Skull not over ⅔ as broad as long. Dentition including enlarged crushing or biting teeth on margins of jaws. Vertebrae without swollen neural arches.
   A. Size small, skull length 16–25 mm. Orbit with raised bony rim.
      1. Skull broad. Maxillary teeth fairly uniform to seventh behind which they suddenly decrease in size. Wichita group—*Pariotichus.*
      2. Skull narrow, coronoids dentigerous; maxillary teeth of uniform height but increasing in diameter to large, compressed teeth below orbit, then abruptly smaller; tips of anterior teeth recurved. Arroyo formation, Clear Fork group—*Cardiocephalus.*
   B. Size larger, skull length over 30 mm. Maxillary and jugal border flared outward. Faint sculpture on dermal bones, no raised orbital rim. Cheek teeth round, rapidly increasing in height posteriorly to enormous tooth below orbit; then abruptly smaller. Short and possibly edentulous coronoid area. Orbital rim not prominent. Arroyo formation, Clear Fork group—*Euryodus.*

II. Larger animals, skull width over ¾ skull length. Vertebrae with massive neural arches and zygapophyses.
  A. Dentition of blunt crushing teeth on palate and coronoids; marginal teeth of jaws not enlarged. Putnam formation, Wichita group, possibly also in Clear Fork—*Pantylus*.
  B. Dentition of numerous slender, pointed teeth along jaw margins. Skull depressed and pointed anteriorly. Arroyo formation, Clear Fork group—*Ostodolepis*.
III. Large. Characters not known. Clear Fork group—*Goniocara*.

STRATIGRAPHIC DISTRIBUTION OF THE GYMNARTHRIDAE

Accurate delimitation of the stratigraphic range of gymnarthrid genera is prevented by the scarcity of remains, which obviously show only a few glimpses of their distribution in time and space, and also by the lack of precise data as to the localities from which certain types were obtained. Pertinent details have already been given for the various species, but a brief summary may be useful. In general, the range of the Family is through the Wichita group and early part (Lower Arroyo formation) of the Clear Fork group of the early Permian in north Texas. This corresponds to the late Wolfcampian and early Leonardian series of the marine Permian standard.

Known occurrence of the various genera is summarized in the accompanying table.

TABLE 4

Stratigraphic distribution of gymnarthrid genera

	Pantylus	*Pariotichus*	*Ostodolepis*	*Goniocara*	*Euryodus*	*Cardiocephalus*
Clear Fork Group						
Choza formation						
Vale formation						
Arroyo formation	(?)		x	x	x	x
Wichita Group						
Leuders formation						
Clyde formation						
Belle Plains formation						
Admiral formation		.				
Putnam formation	x					
Moran formation						

Records marked ? are doubtful as to locality or formation; those marked (?) are of questionable generic reference.

THE AUDITORY REGION OF MICROSAURS

One side of the paratype skull of "*Gymnarthrus willoughbyi* Case," A.M.N.H. no. 4763a (fig. 13), reveals the external form of the auditory region quite clearly and affords an opportunity for comparison of the microsaur ear with that of other primitive tetrapods. These structures are seen in the illustrations of the palatal and occipital surfaces of the skull (figs. 5 and 13).

The stapes of *Cardiocephalus* consists of a large, oval, outwardly convex footplate, which forms the posteroinferior corner of the braincase, and a short, round shaft which extends forward and outward in the direction of the thickened mid-section of the quadrate. There is no trace of a stapedial foramen or of a dorsal process. A mold of the inner surface of the right stapedial footplate shows that it was evenly concave. Dorsally the footplate is nearly vertical on the occipital surface and curves around almost into the plane of the palate below. Its anteroventral edge is thickened adjacent to the pterygoid.

A large fenestra ovalis is bounded above and behind by the overhanging edge of the supratemporal, medially by the opisthotic, and ventrally by the expanded posterior plate of the parasphenoid. Laterally its boundaries are concealed on the right side; on the left, where the stapes is absent, traces of bone which may represent the proötic may be seen between the fenestra ovalis and quadrate, but nothing can be determined about their form or relationships.

On the right side of the specimen, above the columellar process of the stapes, and behind the quadrate, lies a small bone whose relationships have not been determined. Inasmuch as the skull roof was crushed inward considerably in the temporal region, it is possible that this represents a portion of the supratemporal or squamosal.

An opening bounded by opisthotic, stapes, and parasphenoid lies at the posteromedial side of the fenestra ovalis and is confluent with it. Structures which might have occupied this unossified space include the vagus (X) nerve and jugular vein; the perilymphatic duct; or an unossified accessory auditory ossicle similar to that termed the operculum in modern Amphibia. The urodele-like stapes and position of the unossified area somewhat posterior to the main fenestra ovalis both favor the latter interpretation.

Other structures of the auditory region may briefly be summarized as follows:

The opisthotic is a trapezoidal bone, its longer base upward, its posterior surface slightly convex. Supratemporal and postparietal overhang its upper edge. Medially its suture with the exoccipital is nearly vertical, passing through the base of the deep pocket into which the hypoglossal foramina open (but not through these foramina). Its lateral edge bounds the fenestra ovalis and is in contact with the stapes throughout. Ventrally it bounds the posterior extension of the fenestra ovalis mentioned above.

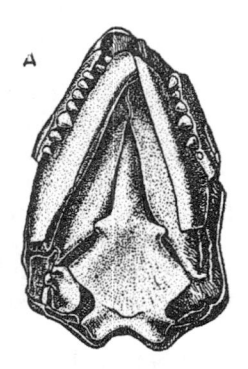

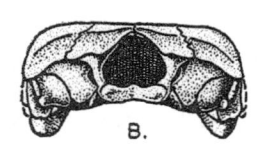

Figure 13. *Cardiocephalus sternbergi* Broili. A.M.N.H. no. 4763a. A. palatal view. B. occipital view showing details of otic region. x 3. Drawn by L. I. Price.

The entire otic region lies below and behind the forwardly inclined quadrate, and is largely concealed from side view in undamaged skulls by the posterior end of the lower jaw and the overlying quadratojugal. From behind or below it is readily visible. On the skull Y.P.M., no. 3689, the occipital surface curves forward toward the side, a flange of the supratemporal (?) lies behind the posterior edge of the squamosal and encloses a space for the depressor mandibuli muscle and tympanic cavity. On skull A.M.N.H. no. 4763a it may readily be seen that the stapes and associated structures lie forward of the occipital surface, behind the quadrate, and within the area which would be covered by squamosal and quadratojugal if those bones had been preserved.

In the absence of a stapedial foramen and dorsal branch, the stapes of *Cardiocephalus* differs strikingly from that of Labyrinthodonts (Sushkin, 1927, p. 281–286; Romer, 1947, p. 62–63) or primitive reptiles (Sushkin, Ibid., 304–307; Price, 1935, p. 382–383, Pls. 7, 9). Its large footplate and rudimentary shaft directed toward the quadrate are similar to the stapes of urodeles, gymnophionans, *Lysorophus,* and the aistopods, i.e., all lepospondylous Amphibia to which microsaurs have been regarded as distantly allied on the basis of vertebral structure. Inasmuch as burrowing lizards of the family Amphisbaenidae have a somewhat similarly shaped stapes (Zangerl, 1944) and further agree with the above lepospondylous Amphibia in the forwardly inclined quadrate and anterior jaw articulation, it cannot be assumed that the resemblances of the auditory apparatus are proof of close affinity.

Cardiocephalus resembles modern urodeles, particularly *Ambystoma* in the form and position of its stapes and the posteroventral enlargement of the fenestra ovalis which may have been covered by a cartilaginous operculum. It differs from *Ambystoma* (as well as from other urodeles) in the relatively larger stapedial footplate in proportion to skull size, and in the contact of the parasphenoid with the rim of the fenestra ovalis. In this latter feature it approaches the Gymnophiona, which also possess an extremely wide parasphenoid. The stapes itself does not show more similarity to the gymnophionans than to urodeles, and appears to cover the fenestra ovalis less completely than that of the modern blind worms in spite of its relatively larger size. The columella of gymnophionans is pierced by the stapedial artery. However the ossified stapes lacks a foramen and presumably the ossification does not extend distally to include that structure. Gadow (1923, p. 85) states that the stapes of *Ichthyophis* is perforate. The fenestra ovalis of *Cardiocephalus* is not strictly lateral in position, as it is in caecilians.

Among fossil Amphibia the resemblance of the large stapes to that of the aistopods is immediately striking; the general relations of the otic region are similar to that group. The stapes of *Phlegethontia* (Gregory, 1948, p. 643) differs from that of *Cardiocephalus* in completely filling the fenestra ovalis, and in having less shaft or columellar process. Possibly the latter feature is merely a difference in ossification. However the fossorial modifications of the *Phlegethontia* skull suggest that the middle ear may have been reduced in aistopods as it is in snakes and salamanders. The posteroventral position of the fenestra ovalis and stapes of *Cardiocephalus* is more like that of *Phlegethontia* than any other comparable form.

So far as we are aware, the stapes has not been described in nectridians.

Lysorophus (Sollas, 1920, fig. 8) has a relatively smaller stapedial plate than *Cardiocephalus*. It is confined to the lateral surface of the skull. Its form is more like that of the amphisbaenids and gymnophionans than the aistopods or *Cardiocephalus*. The reduced skull roof and large supratemporal of *Lysorophus* has led Romer (1945, p. 160, 592) to place it with the microsaurs; Watson also associated it with more typical microsaurs in his Order Adelospondyli (1929, p. 245). But the unique vertebral structure and skull roofing pattern, and the different development of the otic region from other microsaurs, still isolate it from the remaining lepospondyls. To what extent the similarities are the result of convergence rather than phyletic affinity cannot be answered until much more is known of the origin and early deployment of the lepospondylous Amphibia.

GYMNARTHRID TOOTH IMPLANTATION AND SUCCESSION

The dentition of gymnarthrids may be illustrated best by comparison of *Cardiocephalus* and *Euryodus*. Certain details are more conveniently studied in the latter because of its larger size. Although these two genera are closely related, a number of differences in the dentition suggests different modes of life.

Marginal teeth of the gymnarthrids are primitively acrodont, and are implanted in shallow, alveolar pits. Occasionally *Euryodus* has a faintly developed, broad, alveolar depression extending the length of its dentary.

Tooth replacement is demonstrated best by several dentaries of *Euryodus*. A developing tooth impinges on the posterolingual aspect of an older tooth causing resorption of the latter and of the adjoining root area. Gradually the new tooth moves into the position of the antecedent tooth. In one large, massive dentary, probably of an old individual (fig. 14), the labial portions

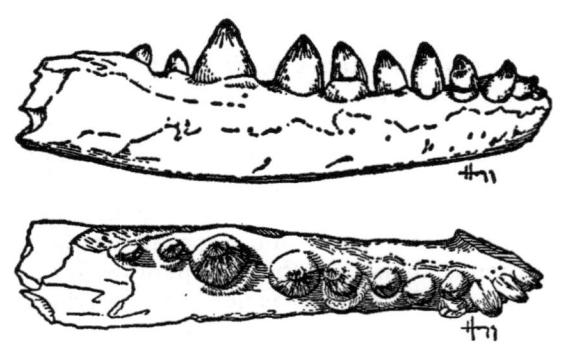

Figure 14. *Euryodus primus* Olson. Lower jaw showing tooth replacement. K.U.M.N.H. no. 9922. x 3. [original 2.4 cm. long overall]. Drawn by Victor Hogg.

of several older teeth may be seen persisting as fragmentary half shells. Normally the older teeth do not endure for so long a time. A definite order of replacement cannot be determined, if there is one. The great majority of the numerous dentaries and maxillaries of both genera show a full tooth row with no indication of a primitive "wave" replacement. Apparently in the relatively short tooth row individual teeth were replaced rapidly. One feature noted is the tendency for the largest cheek tooth to shift backward as the tooth row lengthens.

Sections through tooth crowns of *Cardiocephalus* and *Euryodus* show no folding of the thin enamel, although the latter forms numerous, minute striae radiating from the tip of the crown. The dentine possesses numerous, minute canaliculi radiating evenly from a relatively large pulp cavity. The shallow, alveolar pits seen in a few dentaries of *Euryodus* have septate divisions around the perimeter suggesting labyrinthine structure at least at the base of the tooth. Thin sections (fig. 15) taken through the base

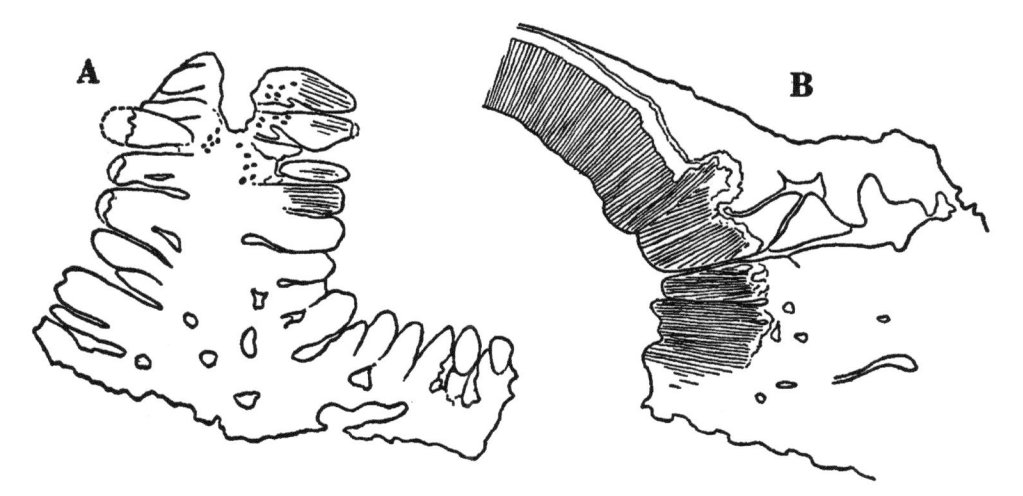

Figure 15. *Euryodus primus* Olson. K.U.M.N.H. no. 9936. Sections of dentary teeth. A. through bases of two adjacent teeth. B. obliquely through tooth near base. x 25. Lobules of dentine well developed at base in A. In B., lobule formation beginning at base involves little of enamel layer. Drawn by F. E. Peabody.

show that the septae are closely packed lobules of dentine in which the canaliculi, as in the crown, remain evenly radiate from the central axis of the tooth rather than orienting with the individual lobules. It is doubtful whether the enamel is involved in the formation of the dentine lobules. In any case, the septate structure, if labyrinthine in origin, appears to be highly specialized, adapted to anchor the tooth, and does not affect the crown of the tooth.

In *Cardiocephalus* the crown of marginal teeth especially in the anterior half of the jaws shows a delicate, sharp keeling and relatively strong recurving at the tip (fig. 8 F). The teeth in the posterior half of the jaw are consistently compressed laterally in contrast with the nearly perfectly rounded crown in *Euryodus*. The coronoid teeth of *Cardiocephalus* (fig. 8 F–G) are arranged in two parallel rows, the labial row consisting of larger teeth. All these teeth are blunt, erect cones, but the small palatal teeth on the middle area of the pterygoid are sharp cones inclined strongly backward. The nature of the palatal and coronoid teeth of *Euryodus* is unknown. Presumably the coronoid dentition, at least, was weaker and more restricted anteriorly than in *Cardiocephalus*, judging from the coronoid suture on the dentary bone.

Occlusion of the jaws brings the upper marginal teeth lateral to the lower teeth with a relatively powerful, trenchant, and crushing action. Available specimens do not show clearly whether or not the coronoid teeth met palatal teeth growing on the ectopterygoid and palatine. Only one or two of all marginal teeth observed show any wear on the crown. One such example of wear in *Euryodus* occurs on a tooth with a resorption pit in its side. Even here the crown tip is only slightly blunted. Evidently constant replacement maintained a sharp dentition. All features of the dentition of *Cardiocephalus* suggest that a trenchant, piercing bite was important in catching and killing animal food. Crushing action was of secondary importance else the pterygoid denticles would not be so greatly inclined rearward. *Euryodus* seems to have had even less crushing action performed by palatal and coronoid teeth, and less trenchant action of marginal teeth. The bulbous but pointed cheek teeth may have been efficient for a piercing, killing bite but obviously included a crushing component.

The distribution of teeth in *Cardiocephalus* and in *Euryodus* illustrates two different adaptive trends in closely related genera. *Cardiocephalus* has what may be considered the more generalized pattern of distribution with main emphasis on marginal teeth but maintaining strong palatal and coronoid accessories. *Euryodus* has a pattern showing greater emphasis on marginal teeth with development of an enlarged, powerful cheek tooth and a powerful "masseter" inserting in a deep fossa on the dentary correlated with a reduced coronoid dentition. Probably the palatal dentition is also reduced, although the lack of teeth here in the type and paratype specimens of *Euryodus primus* may be the result of poor preservation. If the controversial *Pantylus* be considered a gymnarthrid, then its dentition represents another trend, away from strong marginal teeth and toward strong emphasis on crushing action by extensively developed palatal and coronoid teeth. In it, the lower marginal dentition, as illustrated by Williston (1925, fig. 18), becomes merely an accessory to the broad field of coronoid teeth.

ISOLATED POSTCRANIAL BONES OF GYMNARTHRIDS

Among the myriads of dissociated bones washed from the matrix of the Fort Sill fissure deposits are many specimens of the delicate bones of gymnarthrid skeletons. Identification of these is difficult, as several types of small tetrapods occur here. When obviously reptilian and labyrinthodont specimens have been eliminated, a set of elements of a single morphological type remains. Some of these, dorsal vertebrae and humeri, can be shown to belong to *Cardiocephalus* by comparison with the articulated skeleton. Probably many if not all the remaining elements pertain to either *Cardiocephalus* or *Euryodus*. A noticeable bimodality in the size distribution of nearly all these bones may indicate mixture of elements of these genera, which at present cannot be distinguished. The larger size gymnarthrid bones are much less common than the smaller group, which would accord with the greater abundance of the small *Cardiocephalus* than of the larger *Euryodus*, although it might also result from selective interment of immature rather than adult individuals.

Why so few vertebrae and girdles occur in comparison to proximal limb bones is not clear. The rarity of the relatively small radii, ulnae, and fibulae, and of the scapulocoracoid in comparison to pelvis, may be due to excessive breakage of more fragile elements to unrecognizable fragments.

OCCIPITO-ATLAS REGION

Details of the occipital region are poorly preserved in known gymnarthrid skulls, but it is clear that the skull articulated with the atlas by means of two distinctly separate condyles. A number of isolated specimens from Fort Sill referred with reasonable certainty to *Euryodus* and *Cardiocephalus* provide excellent detail of this region.

The exoccipitals of *Cardiocephalus* (fig. 16, E–G) each provide a distinct condyle with a flattened articular surface in the same plane as its fellow, and the plane is inclined anteroventrally 30° with the horizontal axis. Between the two condyles lies a deep and broadly concave pit in a well developed basioccipital bone. The pit faces posterodorsally, approximately at a right angle to the plane of the occipital facets. The whole arrangement of articular surfaces is very close to that of *Cryptobranchus*, except that in the latter there is no definitive[*] basioccipital and the enlarged parasphenoid extends back to form the ventral wall of the pit. Also in the urodele, the plane of the one articular surface on the exoccipital extends anteromedially

[*] But a large skull of *Cryptobranchus* at hand has a thin bony plate wedged between the dorsomedian edges of the basal plate of the exoccipitals in the position of a vestige of the basioccipital.

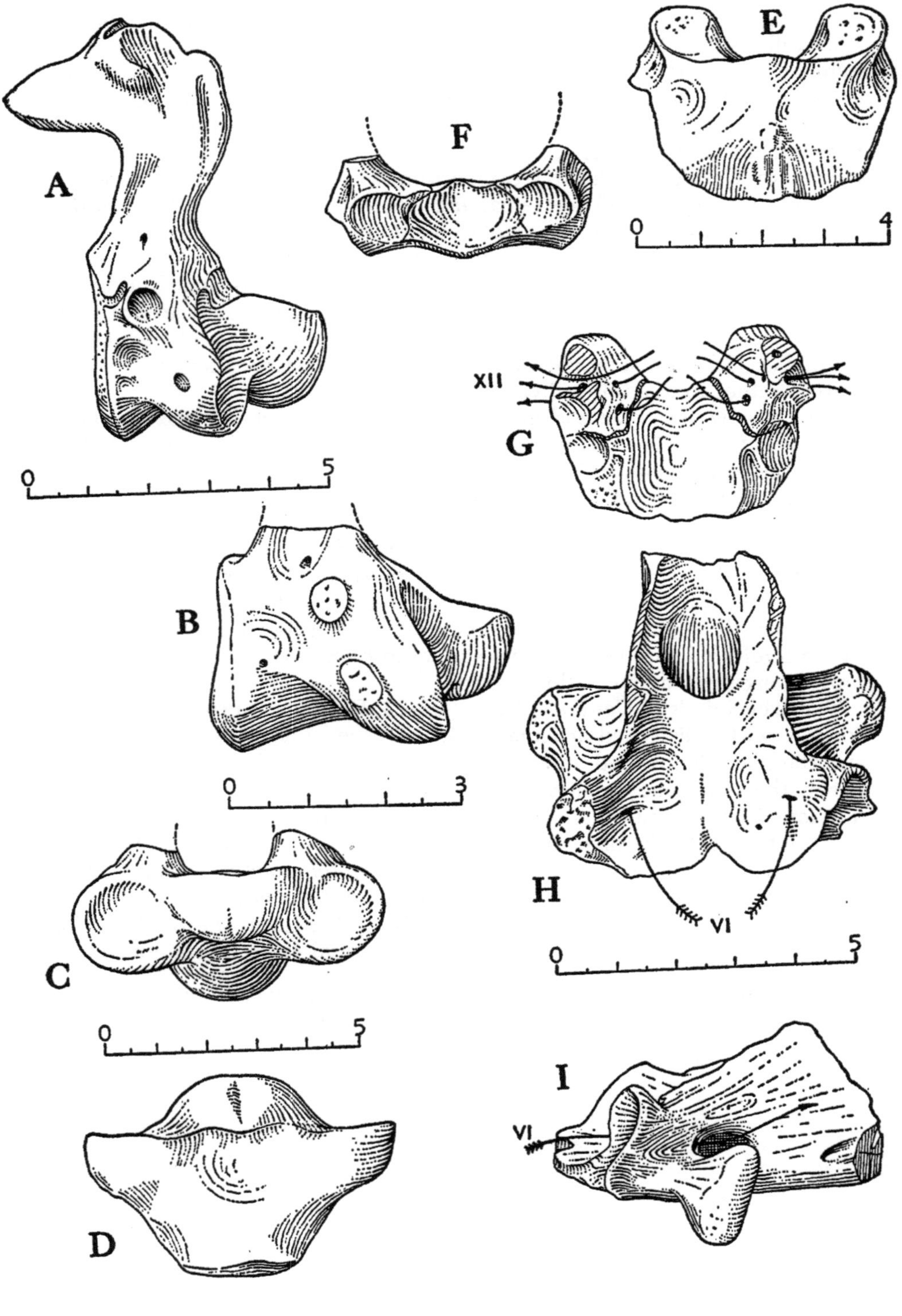
A
0 5
F
E
0 4
XII
G
B
0 3
H
VI
0 5
C
0 5
D
I
VI

rather than in the same plane as the corresponding surface on the opposite condyle. Lateral articular surfaces on the atlas are correspondingly different. The suture with the basioccipital is clearly seen on the dorsal, posterior and lateral aspect, but is invisible ventrally, thus explaining why it is difficult to interpret this region in gymnarthrid skulls. The basioccipital has a finished dorsal surface (according to Romer, 1947, p. 49, never found in labyrinthodonts on this element) which is deeply concave; ventrally a distinct overlap by the parasphenoid is indicated by a recessed shelf; laterally a spherical depression in the sutural surface presented to the opisthotic represents a part of the otic labyrinth.

Three foramina (fig. 16 G) are clearly evident on the inner aspect of the exoccipital. The most anterior one is ventralmost in position and opens low on the lateral surface. The most posterior foramen is dorsalmost and the largest; the third foramen seems to be a branch of the latter. There is little doubt that the foramina are for branches of a hypoglossal nerve and so constitute a primitive character lost in advanced apsidospondylous and lepospondylous amphibians but retained in the amniote line.

Material demonstrating the gymnarthrid atlas consists of: a nearly perfect atlas of *Euryodus* (fig. 16 A) associated with a fragmentary second cervical and showing no evidence of intercentra; a centrum with the bases of atlas arches present (fig. 16 B–D) and two fragments (K.U.M.N.H. no. 9927) representing *Cardiocephalus*. The atlas of *Euryodus* is approximately the same size as that of a *Necturus* having a skull length of 35 mm.

The atlas is differentiated from a lepospondylous design and closely resembles the atlas of *Cryptobranchus* except for the slightly broader "odontoid" knob fitting into the basioccipital depression and higher dimensions overall correlated with a slightly higher skull. The centrum is deeply concave and notochordal in rear view, but the "odontoid" has only a slight dorsoventral indentation probably not related to a notochordal depression, and there is no notochordal pit, as such, in the basioccipital bone. Remains of a suture between centrum and arch are clearly evident in *Euryodus* (fig. 16 A) especially on the median side of the arch. The pedicel of the arch bears a circular diapophysis distinctly separate from a similar parapophysis placed lateroventrally on the centrum. Clearly the atlas bore a two-headed rib. The arch is pierced a short distance above the diapophysis by a tiny foramen for the exit of a cervical nerve. The contours of the arch are simple in *Cardiocephalus*, but in the larger *Euryodus* (fig. 16 A) the crown of the arch exhibits a transverse, chevron-like furrow indicating a cartilaginous process. Anterolaterally from the end of the furrow there

---

Figure 16. Isolated gymnarthrid bones. *Euryodus primus* Olson. A–D, H, I; *Cardiocephalus* cf. *sternbergi*. E–G. A, Atlas, lateral view, C.N.H.M. no. UR 287. B–D, atlas, lateral, anterior, and ventral views, K.U.M.N.H. no. 9926. E–G, exoccipital-basioccipital complex, ventral, posterior, and dorsal views, K.U.M.N.H. no. 9928. H–I, basisphenoid-parasphenoid complex, dorsal and left lateral views, C.N.H.M. no. UR 286. Scale in millimeters. Drawn by F. E. Peabody.

extends a prominent ridge probably for muscle attachment. The excellent detail here indicates no surface which could be a prezygapophysis articulating with a proatlas.

The occipito-atlas articulation of gymnarthrids appears to be a strong interlocking structure commensurate with strong jaws and dentition; it is definitely lepospondylous and urodelian in design. It would be difficult to distinguish the atlas of gymnarthrids from a urodele such as *Cryptobranchus*, indeed the whole occipito-atlas complex of the latter could easily have evolved from the gymnarthrid structure by degeneration of the basioccipital, by compensatory backward growth of the parasphenoid, and by loss from the exoccipital of foramina for branches of the hypoglossal nerve.

DORSAL VERTEBRAE: All presacral vertebrae except the atlas are similar in form (figs. 17, 18). Slender neural arches bearing unswollen zygapoph-

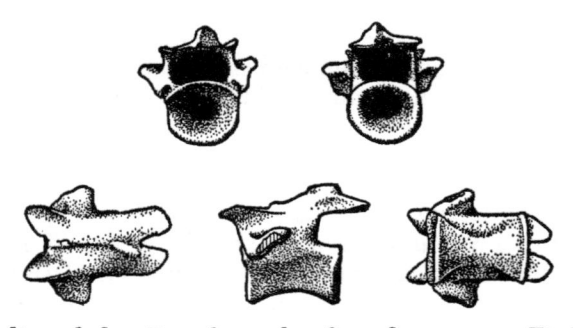

Figure 17. *Cardiocephalus*. Dorsal vertebra from fissures near Ft. Sill, Oklahoma. Y.P.M. no. 3819, dorsal, lateral, ventral, anterior, and posterior surfaces. x 5.

yses are united to the hourglass-shaped notochordal centrum by persistent sutures. The centrum is short and wide, its ends parallel and vertical. The articular faces are circular and pierced by deep conical excavations which meet to form a canal for the persistent notochord. Slight dorsolateral extensions of the articular surfaces near the base of the neural arch pedicels form facets for the rib capitulum. The lower surfaces of the centra are round, smooth, evenly constricted toward the middle of the vertebrae. Degree of constriction varies from slight concavity of the sides and base to strongly hourglass-shaped, but is always gradual, never an abrupt pinching in of the sides such as one finds in *Captorhinus*. A small nutrient foramen may be present on either or both sides of the centrum near the middle.

Above the center, the sides of the centrum flare out to support the base of the neural arch. The dorsal surface of the centrum forms the floor of the neural canal. Here the hourglass form of the centrum is clearly visible inside the arch pedicels, the canal following its contour and hence deepest near the middle of the vertebra.

The neural arch is suturally united to the centrum, the condition typical of microsaurs and termed adelospondylous by Watson. The suture occupies the anterior three-fourths of the length of the centrum, and is wedge shaped, descending from the anterior end to an angle just in front of the middle of the vertebra, and then rising posteriorly.

Transverse processes arise near the anterior end of the neural arch pedicels. They slope downward and forward, and are widest at their posterior border. They somewhat resemble those of anterior dorsal vertebrae of *Varanosaurus* (Williston, 1911, Pl. 2), but are shorter than most pelycosaur diapophyses. Diapophyses of the anterior vertebrae of *Captorhinus* are of

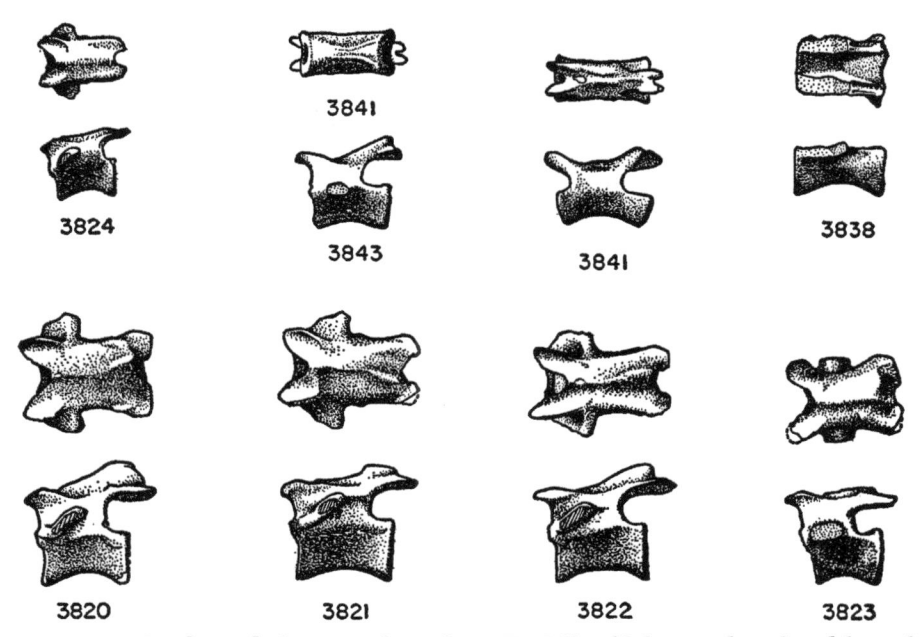

Figure 18. *Cardiocephalus*. Vertebrae from Ft. Sill, Oklahoma, dorsal and lateral views. Y.P.M. nos. 3820, 3821, 3822, 3824, presacral. Note relatively tall neural spine of no. 3820. Y.P.M. no. 3838 centrum of immature dorsal showing neural suture. Y.P.M. no. 3823 sacral vertebra. Y.P.M. no. 3841 caudal vertebra *Cardiocephalus* or ?*Captorhinus*. Y.P.M. no. 3843 ?*Cardiocephalus* proximal caudal or sacral. All x 5.

uniform width and extend farther ventrally, onto the centrum, providing an appreciably longer articular facet for the rib. Little variation in shape or size of the diapophyses exists between the various vertebrae of *Cardiocephalus,* in contrast to the marked regional variation of these processes in reptile vertebrae. Unlike typical reptiles, there are no cervical vertebrae with low parapophyses and downwardly directed diapophyses; the articular facets for ribs are confluent on all observed specimens.

The roof of the delicate, low, flat-topped neural arch slopes forward slightly from above the postzygapophyses to the prezygapophyses. It meets the lateral walls of the pedicels at nearly a right angle along ridges prolonging the edges of the zygapophyses. The arch is slightly wider than the ends of the centrum and fairly uniform in width throughout. It extensively overlaps the anterior end of the following vertebra, completely covering the spinal cord. A faint rounded median ridge is present for most of its length; this may rise to a weak neural spine just anterior to the posterior zygapophyses on more posterior (?) dorsals. Frequently small irregular tubercles for attachments of ligaments are present above the posterior zygapophyses.

SACRAL VERTEBRA: A few vertebrae similar in structure to those just described but bearing a large oval, horizontally elongated rib facet low on the side of the centrum are regarded as sacral vertebrae of *Cardiocephalus* (fig. 18, Y.P.M. no. 3823). The transverse process is formed entirely by the neural arch, although its top lies at the level of the floor of the neural canal; the suture between arch and centrum extend lower on these vertebrae. The sides of the centrum are excavated so that a pair of rounded ridges lies on either side of the midline. Immature centra from which the arches have separated show a deep V-shaped suture, its surface turned outward as well as upward (fig. 18, Y.P.M. no. 3838).

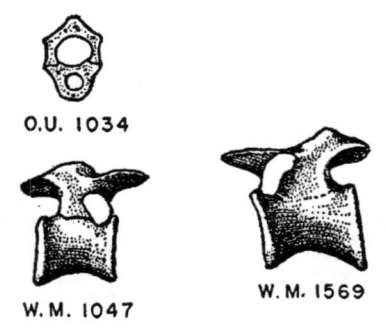

Figure 19. *Cardiocephalus*. O.U. no. 1034, natural cross section of posterior dorsal vertebra. W.M. no. 1047 lateral view. *Euryodus*, W.M. no. 1569. Dorsal vertebra. All x 4. Drawn by L. I. Price.

Several sacral vertebrae in the Peabody Museum collection are larger than the majority of the *Cardiocephalus* dorsals (fig. 20 A). They agree with *Cardiocephalus* rather than *Captorhinus* in the rounded lower surface of the centrum but differ from the sacral just described in lacking the lateral excavation. Posterior zygapophyses are a little smaller, but do not differ much from those of anterior caudal vertebrae of *Captorhinus*. Some, but not all, of the latter show traces of hyposphenes. Hence it is not certain that these specimens (nos. 3851, 3852, 3853) are not immature *Captorhinus* vertebrae.

CAUDAL VERTEBRAE: A number of small caudal vertebrae have been referred to *Cardiocephalus* with great hesitation (fig. 18, Y.P.M. no. 3841). They resemble those of *Captorhinus* in the inverted trapezoidal shape of the ends of the centra, in the convergence of the planes of the end of the centra ventrally, and in the presence of a pair of distinct keels or ridges along the lower edge of the centrum which converge in the middle but diverge toward the ventrolateral corners of the articular discs. The neural arch is slender, and narrower than the centrum; the zygapophyses are close together, and slope inward. A low spine rises over the postzygapophyses. The only features which distinguish these from known caudal vertebrae of *Captorhinus* are their smaller size and the absence of any trace of the transverse fissure permitting autotomy of the tail, described in *Captorhinus* by Price (1940) and well defined in material at hand. Possibly these are distal caudals of *Captorhinus* which did not develop the transverse fissure. If so, no caudals of *Cardiocephalus* have been recognized.

COMPARISONS: Dorsal vertebrae of *Cardiocephalus* resemble those of such typical microsaurs as *Microbrachis pelikani* Fritsch and *Hyloplesion longicaudatum* Fritsch in the hourglass-shaped centra, presence of a suture between neural arch and centrum, and broad zygapophyses. Steen (1938, p. 230, p. 233, fig. 19C) states that the centrum of *Microbrachis* carries a large facet for the capitulum of the rib near its anterior end. In *Cardiocephalus* the rib attaches intervertebrally and the facets are larger at the posterior than anterior ends of the centra. From both *Microbrachis* (cf. *M. mollis* Fritsch, 1879, p. 180, fig. 116) and *Hyloplesion* (*Ibid.*, Pl. 39, fig. 3) *Cardiocephalus* differs in the absence of even a low neural spine.

The vertebrae resemble those of *Ostodolepis* (Williston, 1913, p. 364; Case, 1929, p. 100) in the short, oblique rib facet of the high transverse process, but differ from that genus in the absence of a neural spine, less massive posterior zygapophyses, and to judge from both Case's and Williston's illustrations much less massive pedicels of the neural arch and relatively larger canal. A further distinction is the round lower profile of the centrum, which in *Ostodolepis* is "nearly square" (Williston, 1913, p. 363). Williston later (1916, p. 174) regarded the vertebrae of *Pantylus* as similar to those of *Ostodolepis*, stating that the arches are in no way expanded and thickened like those of other cotylosaurs, but rather resemble those of pelycosaurs. His illustration of the caudal vertebrae reveals a more prominent neural spine than is found on the caudals questionably attributed to *Cardiocephalus;* the size, of course, is appreciably greater.

Vertebrae of nectridians differ greatly from those of microsaurs in their elongate neural and haemal spines, and in the nature of their transverse processes which arise near the middle of the vertebrae rather than at the anterior end, and commonly bear two articular facets for ribs. Aistopod vertebrae resemble microsaurs in the absence of a neural spine and wide neural arches, but have centrally located diminutive transverse processes more like those of the nectridians. Vertebrae of *Cardiocephalus* most closely resemble those of the Gymnophiona among recent Amphibia. They differ from those of urodeles in position of the transverse process, which is centrally located as in nectridians in salamander vertebrae.

Cardiocephalus vertebrae may be distinguished from those of *Captorhinus* (with which they are associated at Fort Sill), by smaller size, and for individuals of the same size, more delicate build of neural arch and zygapophyses. They lack the depressed attachment area for an interspinous ligament on the neural arch between the prezygapophyses, and the vestigial hyposphene and hypantrum found on most *Captorhinus* vertebrae (fig. 20 B). Neural arches are relatively longer and narrower than in *Captorhinus*, and the centrum is much narrower at the middle than at the articular ends, yet remains round in cross section, in contrast to the flat bottomed angular profile of *Captorhinus* vertebrae with slightly constricted centra. If caudal vertebrae have been correctly associated, they differ from *Captorhinus* in lacking the groove for autotomy of the tail.

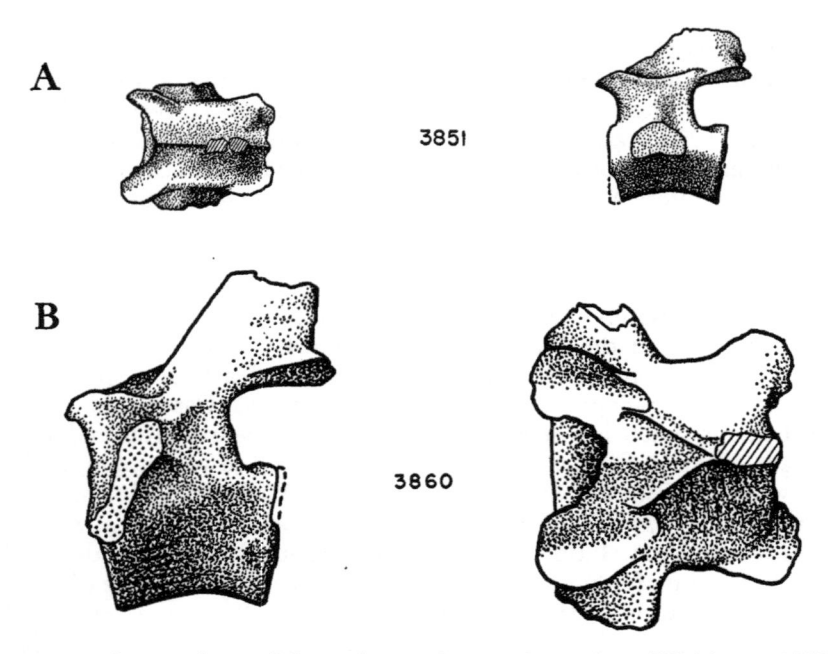

Figure 20. ?*Euryodus* or ?*Captorhinus*. A. sacral vertebra, Y.P.M. no. 3851. x 5. B. *Captorhinus* sp., Y.P.M. no. 3860. Dorsal vertebra, showing relatively massive zygapophyses, more elongate transverse process, for comparison with *Cardiocephalus*. x 5.

The vertebrae differ from those of *Araeoscelis* (Williston, 1914, p. 121, fig. 4) in the more complete and broader roof to the neural canal between the zygapophyses. The zygapophyses do not appear X-shaped in dorsal view, and there is less swelling over the postzygapophyses. Other differences are the confluence of rib facets on the transverse processes, absence of marked excavation of the side of the neural arch, less developed neural spine, and of course the absence of presacral intercentra and of elongate cervical vertebrae. *Tomicosaurus* (Case, 1907, Pl. 27, fig. 8) vertebrae likewise differ in having the zygapophyses much wider than the intermediate portion of the neural arch.

Bolosaurus has vertebrae of similar size, but with much more massive buttresses to the zygapophyses and a relatively tall neural spine.

TABLE 5

Measurements in millimeters of vertebrae

	Number	Observed range	Mean	Standard deviation	Coefficient of variability
Dorsal vertebrae					
Length of centrum	22	1.59–2.87	2.326 ± 0.073	0.343 ± 0.052	14.76 ± 2.23
Width anterior end centrum	22	1.11–2.24	1.846 ± 0.064	0.299 ± 0.045	16.18 ± 2.44
Sacral vertebrae					
Length of centrum	8	1.55–2.50	1.809 ± 0.106	0.300 ± 0.075	16.57 ± 4.14
Width anterior end centrum	8	1.23–2.01	1.660 ± 0.081	0.229 ± 0.057	13.78 ± 3.44
Large sacrals, possibly *Euryodus*					
Length of centrum	3	2.84–4.87	3.58		
Width of centrum	3	2.98–4.06	3.58		
Caudal vertebrae					
Length of centrum	6	2.00–2.97	2.500 ± 0.130	0.318 ± 0.092	12.70 ± 3.66
Width anterior end centrum	5	1.18–1.54	1.346 ± 0.056	0.125 ± 0.40	9.32 ± 2.95

RIBS

Slender, curved, two-headed ribs are present on the exposed portion of the vertebral column of O.U. 1034, and presumably ribs of this type extended from the second vertebra to the pelvis, and perhaps accompanied the anterior caudal vertebrae. No direct evidence is available as to the presence of caudal chevrons or the association of the numerous tiny chevrons recovered from the fissure matrix with this or another genus. Nor has any attempt been made to associate the small two-headed ribs in that deposit with any particular reptile or amphibian.

Two tiny sacral ribs in the Kansas University collection are distinctly different from those of *Captorhinus* and in all probability belong to *Cardiocephalus*. Tuberculum and capitulum are distinct. The short shaft is somewhat twisted and the distal surface is cupped, dorsally convex (fig. 22). It would readily fit the curved iliac blade of the pelves described below.

Shoulder Girdle

Several minute, perfectly formed scapulocoracoids presumably belong to gymnarthrids. They consist of a broad, low scapular blade with a rounded

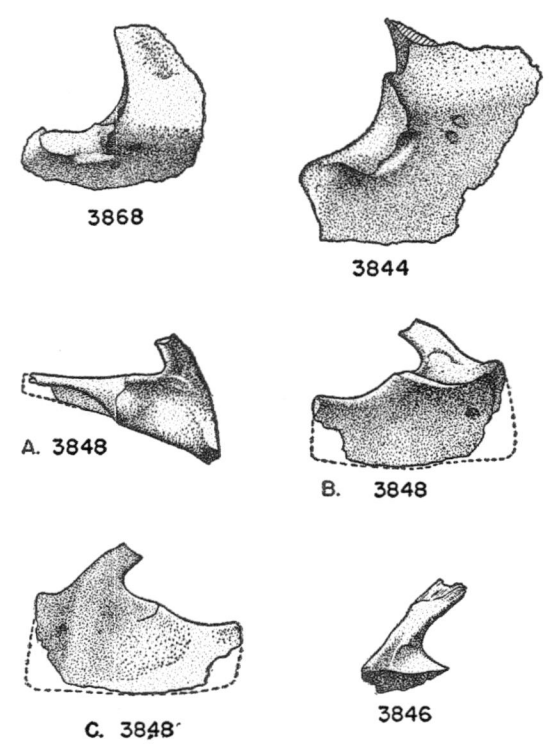

Figure 21. *Cardiocephalus*. Scapulocoracoid, Ft. Sill, Oklahoma. Y.P.M. nos. 3868 and 3844 ventrolateral views. x 5. *Cardiocephalus*. Pelvis. A. right innominate, Y.P.M. no. 3848, lateral view; pubis partially restored from no. 3850. B. ventral aspect of same. C. dorsal surface of same. Y.P.M. no. 3846 immature ilium. All x 5.

anterodorsal edge along which the bone is "incomplete" and was continued by cartilage, and an inwardly turned coracoid portion which bears the "screw-shaped" glenoid fossa. The coracoid region is narrower than the scapula; in immature specimens such as K.U.M.N.H. no. 9929 (fig. 22) and Y.P.M. no. 3845 it is extremely short, scarcely completing the glenoid, and obviously was continued both medially and posteriorly by cartilage. Larger specimens [Y.P.M. nos. 3844 and 3868 (fig. 21) and K.U.M.N.H. no. 9930 (fig. 22)] have the posterior end of the coracoid region complete. Its border descends vertically (actually medially) a short distance behind the glenoid, meeting the median edge of the bone. On these presumably adult girdles the scapular blade lies entirely anterior to the glenoid, and the coracoid projects backward from the front of the glenoid a distance equal to the height of the scapula above that point. A prominent posteriorly facing

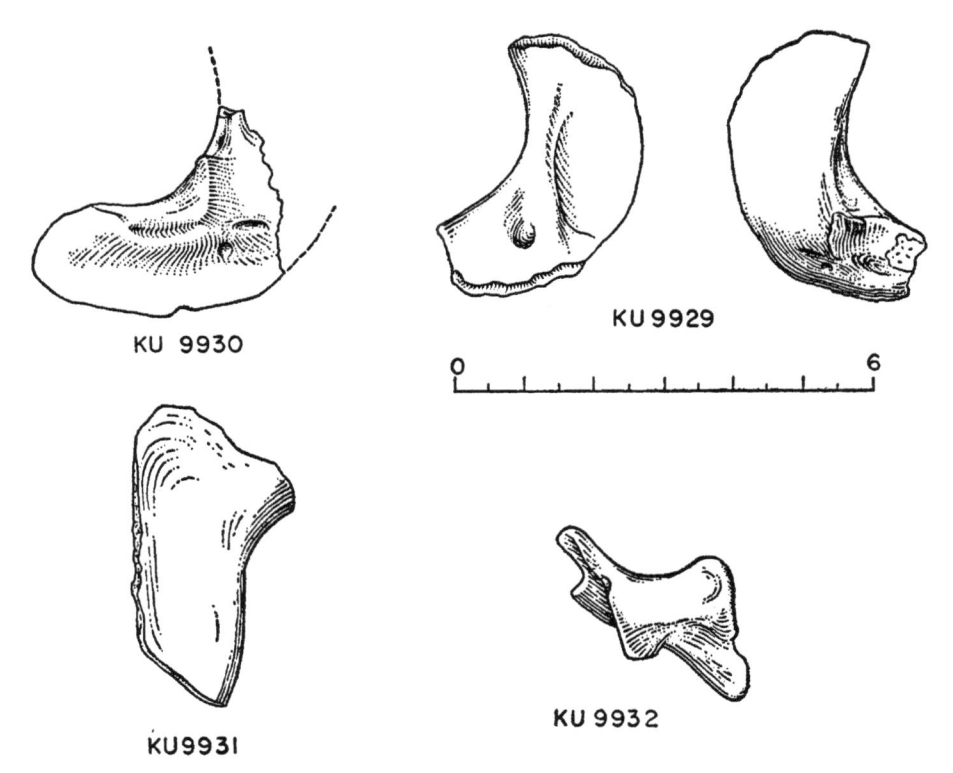

Figure 22. *Cardiocephalus.* Scapulocoracoid. K.U.M.N.H. no. 9930, lateral view, and K.U.M.N.H. no. 9929 internal and lateral faces. Sacral rib, K.U.M.N.H. no. 9932. Ischium, K.U.M.N.H. no. 9931. x 10. Drawn by F. E. Peabody.

supraglenoid buttress area lies internally to the glenoid and is pierced by a slit-like supraglenoid foramen. The oval coracoid foramen lies forward and below the anterior end of the glenoid. Both these foramina open internally in the deep subscapular fossa in front of the supraglenoid buttress. A smaller, round glenoid foramen pierces the coracoid region slightly below and behind the coracoid foramen; internally it opens on the flat inner surface of the coracoid plate. Glenoid and coracoid foramina are about the same distance apart on both small and large specimens. The scapular region is thin and slightly convex anteriorly.

TABLE 6

Measurements in millimeters of scapulocoracoid

	Y.P.M. No. 3868
Front scapular blade to posterior end coracoid	4.46
Front scapular blade to anterior end glenoid fossa	2.87
Height scapular blade from anterior end glenoid	2.89
Height scapular blade from lower edge coracoid	4.74

These fragments are of such small size that they may reasonably be associated with *Cardiocephalus*. Many primitive features proclaim amphibian rather than reptilian affinities. Positive identification is not at present possible, but reference to this microsaur seems a reasonable possibility. Steen (1938, p. 233) figures a scapulocoracoid of *Microbrachis* strikingly different in shape from these specimens, which appear closer to the pattern of primitive labyrinthodonts and reptiles.

No remains of the characteristic, broad, long-stemmed interclavicle of microsaurs have been found in the Fort Sill deposit, so this element remains unidentified. Numerous broken clavicles with abruptly widening ventral plates and narrow ascending processes occur; most are too large to associate with *Cardiocephalus,* and no morphological distinctions have been found between these and the smaller specimens. Accordingly nothing can be said of the dermal shoulder girdle.

Pelvis

Over 40 pelvic fragments from the Fort Sill locality are of small size (compared to the more abundant *Captorhinus*) and might possibly belong to *Cardiocephalus*. In favor of such association are the numerous small limb bones which are regarded as remains of this genus. Both shape and proportions of these pelves so closely resemble the considerably larger specimens of *Captorhinus,* that the hypothesis that they belong to immature individuals of that genus requires serious consideration.

Twenty-eight minute ilia appear to belong to the same kind of animal. The iliac spine is slender and curves backward slightly above the acetabulum. Some spines remain slender and taper slightly distally, others become more compressed and slightly broader near the distal end. Both variations can be matched among the larger ilia of *Captorhinus*. Ventrally these ilia terminate in a flat surface which cuts horizontally across the acetabulum and slopes inward and downward to an apex where it intersects the medial buttress of the pelvic plate. Except near the posterior end of the ilioischiadic suture this plane obviously breaks across the internal trabeculae of the bone, and does not follow a suture.

Five innominate bones are fairly complete except for the upper portion of the iliac spine, which is known from isolated ilia. They consist of a flat ventral plate (figs. 21, 23), near the anterior end of which the acetabulum rises, surmounted by a slender, backward curving iliac spine. The ventral surface is flat or slightly concave from side to side. The lateral border of the ischium is concave, swinging outward and forward to a tubercle at the posterior edge of the acetabulum. In front of this is a shallow embayment in the lateral margin of the ventral plate where it intersects the low-lying acetabulum. A small round pubic foramen pierces the plate near its front

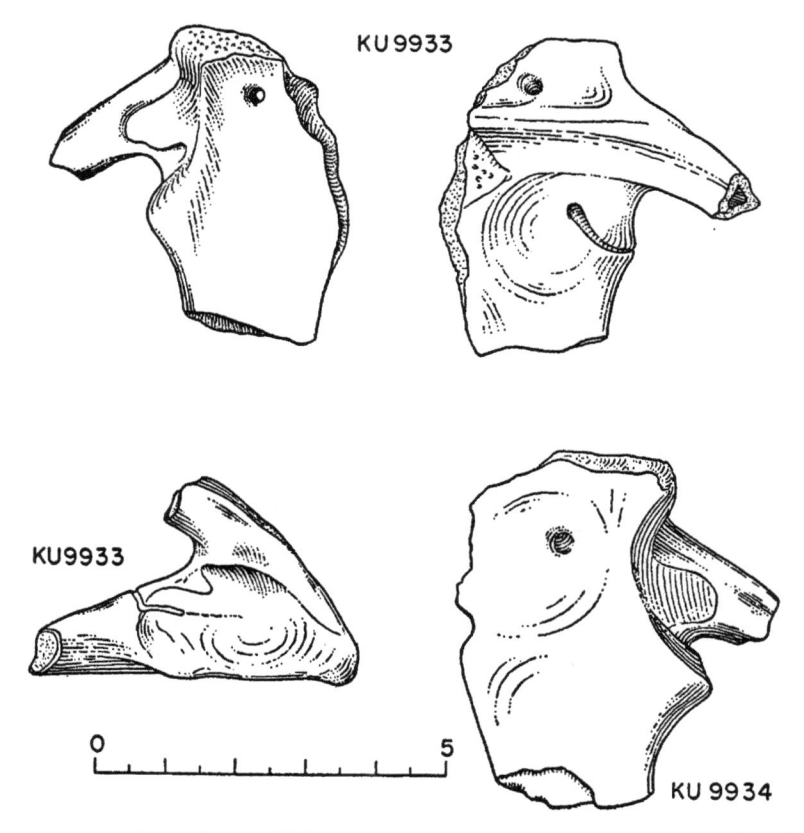

Figure 23. *Cardiocephalus*. Pelvis. K.U.M.N.H. no. 9933, ventral, dorsal, and lateral views. K.U.M.N.H. no. 9934, ventral surface. x 10. Drawn by F. E. Peabody.

end, opposite the anterior half of the acetabulum, slightly closer to the lateral than median edge. Y.P.M. no. 3850 and K.U.M.N.H. no. 9934 show the anterior edge of the pubic region truncated close in front of the corner of the acetabulum, by a surface suggesting cartilaginous extension.

A relatively large acetabulum rises at right angles to the ventral face of the anterior half of the plate. It is situated so low that it intersects the lower surface of the bone. Above it the slender iliac process curves posteriorly. Several specimens have the shaft of the ilium abruptly constricted, rising from the broader base of the acetabulum. Others have a distinctly wider shaft which merges with the contours of the rest of the bone. Perhaps these variations belong to *Euryodus* and *Cardiocephalus,* but there is no association to establish generic identification.

These small ilia and pelves are believed to belong to *Cardiocephalus* rather than *Captorhinus* for the following reasons:

1. Small size and absence of intermediates in size between them and undoubted specimens of *Captorhinus.*
2. Absence of sutures between ilium, ischium, and pubis (aside from an incomplete ilioischiadic suture). Much larger pelves of *Captorhinus* show three well marked sutures, and specimens this small have separate pelvic elements which bear "unfinished' edges for cartilaginous extensions adjacent to the sutures. In particular, young *Captorhinus* pelves have a distinct, discoid pubis. These bones on the contrary suggest that the pelvis was undivided from the start, an amphibian characteristic.
3. They differ from *Captorhinus* in the absence of the ridge on the ventral surface of the pubis which runs anteromedially from below the obturator foramen to the anterior edge of the bone.
4. The medial buttress of the acetabulum is weaker than that of *Captorhinus* and the fossa behind it is shallower than in that genus.

In the University of Kansas collections are three small pelves, seemingly captorhinid and the same size as immature *Captorhinus* specimens, but with no sign of a separate pubis. Possibly these belong to *Euryodus,* whose pelvis has not been recognized otherwise.

Pelvic remains of *Cardiocephalus* are so fragmentary that logical measurements cannot be made on them. An indication of size may be obtained from the illustrations (which are 5 × natural size) and the fact that the incomplete innominate described above (Y.P.M. no. 3848) is 5.36 mm. long and 2.76 mm. wide from midline to lateral edge of the acetabulum. An isolated ischium, K.U.M.N.H. no. 9931 (fig. 22), is 4.5 mm. long. One of the longer ilia measures 4.05 mm. from the front edge of the acetabulum to tip of the preserved portion of the spine. The spine is 0.81 mm. in diameter.

Steen (1938, p. 233, fig. 19A) reports an ossified ilium and ischium in *Microbrachis* but assumes the pubis was unossified. Recent amphibians of course lack an ossified pubis, and it is possible that some microsaurs had already lost that bone. Recalling, however, that primitively the tetrapod pelvis was a single ossification to which later iliac and pubic centers were added, one may speculate that the pubis as a distinct bone ossifying from a separate center never existed in the Amphibia. Presumably the cartilaginous pelvic plate was similar in external morphology to that of the reptile, in which separate ossifications occur. It would be possible for ossification from the ischiadic center to extend forward into the posterior part of the pubic region, as appears to have happened in the specimens here described, and produce a bone closely similar in morphology to the combined pubis and ischium of reptiles, but lacking the suture between these elements.

LIMB BONES

HUMERUS: An incomplete humerus on the articulated *Cardiocephalus* skeleton (O.N. no. 1034) is similar to numerous small humeri washed free from matrix of the Dolese Quarry fissures. These bones (fig. 24) differ from associated *Captorhinus* humeri in smaller size, much more slender proportions, much less expanded proximal and distal ends, relatively more shaft, and in having no entepicondylar foramen. The proximal end is rather similar to that of *Captorhinus*, with spiral head articular area and the deltoid process rather distant from head but in the same plane. The distal end differs from that of *Captorhinus* in lacking the wide and distally projecting entepicondyle. There is no trace of foramen or groove on either epicondyle. Large, well ossified specimens have a well developed spherical radial condyle; others have this less perfectly ossified, and small specimens are hollowed out at both ends where cartilaginous terminations were present in life. The form of these bones is well shown in figure 25, and may be compared with the humerus of *Captorhinus* illustrated in figure 26.

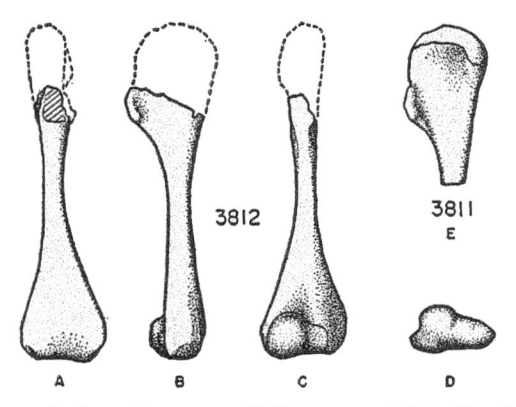

Figure 24. *Cardiocephalus*. Humerus, Y.P.M. no. 3812, Ft. Sill, Oklahoma. A. posterior, B. medial, C. anterior, D. distal surfaces, E. Y.P.M. no. 3811, medial surface. All x 5.

The form of the humerus is not unlike that of some salamanders, but is more primitive in the 90° angle between the planes of the proximal and distal expansions. Also the head is less sharply set off from the proximal muscular crests and is screw shaped rather than subspherical in shape.

Over 130 humeri were recovered. A majority of these are broken across the slender waist; only complete specimens were measured. Measurements

56   REVISION OF THE GYMNARTHRIDAE

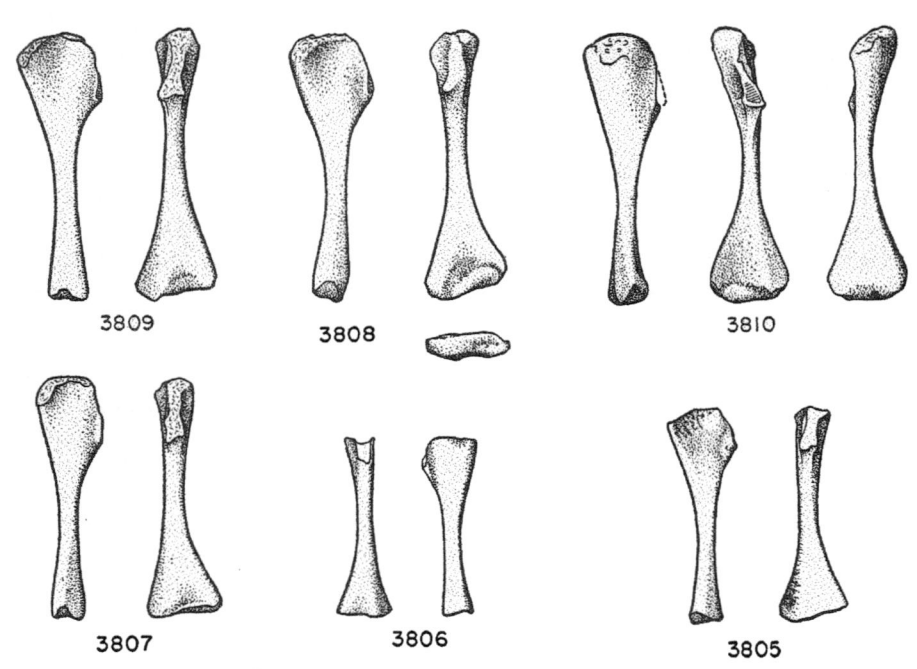

Figure 25. *Cardiocephalus*. Ft. Sill, Oklahoma. Humeri in Y.P.M. collections. All x 5.

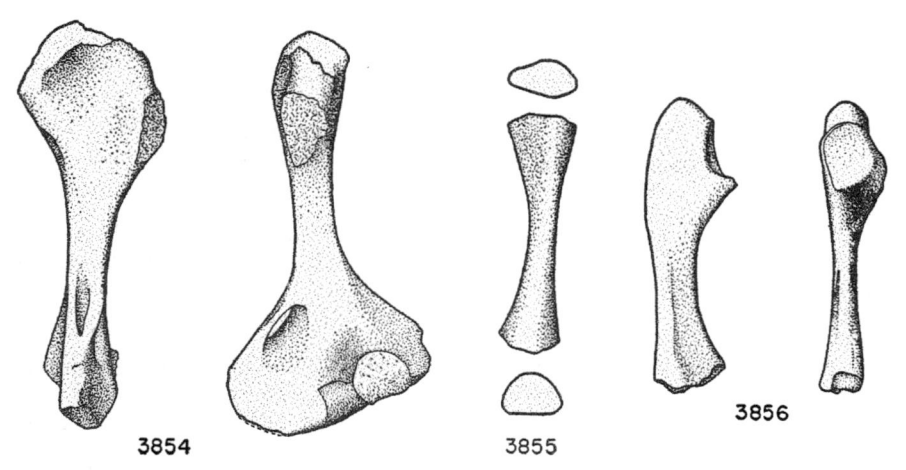

Figure 26. *Captorhinus* sp. Humerus. Y.P.M. no. 3854. Medial and anterior surfaces. Radius, Y.P.M. no. 3855 and ulna, no. 3856, Ft. Sill, Oklahoma. x 2.

were made on the mechanical stage of a microscope with a micrometer reading to 0.01 mm. Length was measured along the long axis of the shaft from points of tangency at head and distal end, the specimen lying on the slide with the proximal expansion flat. Width of head was measured similarly between tangents to the dorsal (extensor) surface and tip of the deltopectoral crest. Width across the epicondyles was taken along a line normal to the length of the bone.

RADIUS: Among the rarest of the limb elements are the slender radii and ulnae whose delicate shafts are extremely liable to damage. In contrast to over 200 femora of *Cardiocephalus* obtained from the Yale sample, only 31 radii and 12 ulnae were recovered.

The radius is typical of that of primitive tetrapods, straight, fairly slender with a distinct waist so that it approaches an hourglass figure in anterior aspect. Its proximal end is compressed, the distal end is almost circular, only slightly flattened posteriorly. Both ends are excavated for cartilaginous extensions similar to the long bones of modern salamanders. Proportions of the radii may be ascertained from figure 27. The length of 25 complete specimens ranges from 2.53 to 4.63 mm., the mean being 3.36 mm.

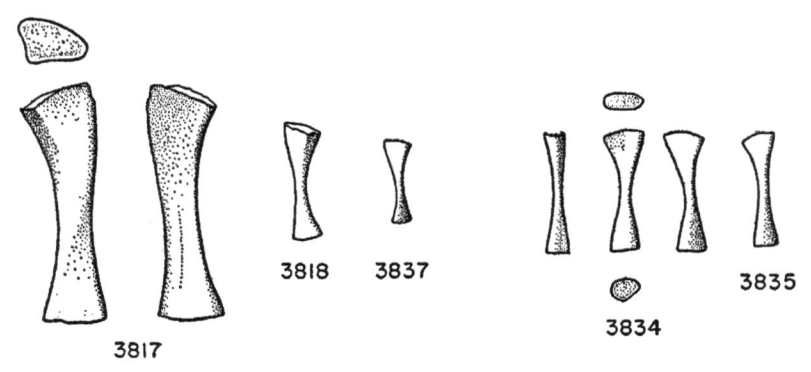

Figure 27. ?*Euryodus* Ulna, Y.P.M. no. 3817, medial and lateral surface and proximal articular face. *Cardiocephalus*. Ulnae, Y.P.M. nos. 3818, 3837; radii, Y.P.M. no. 3834, medial, posterior, anterior, proximal and distal views; Y.P.M. no. 3835, anterior view. x 5.

Three radii greatly exceed the others in size, and have been excluded from the tabulated measurements on this account; their lengths are 6.04, 6.43, and 7.15 mm., respectively. Although these display no morphological features other than large size by which they may be separated from the remainder of the specimens, the presence of other small Amphibia (e.g., related genus *Euryodus*) in the deposit raises doubts as to their association. The smallest of these bones is 5.9 standard deviations from the mean of the remaining sample, a highly significant difference. If included with the rest, the sample has a definitely bimodal distribution and its coefficient of variability is increased from 13.54 ± 1.92 to 29.27 ± 3.91. The former value is comparable to that for other limb bones of *Cardiocephalus*.

ULNA: The minute ulnae are somewhat difficult to distinguish from radii, but have a relatively bulkier head, roughly square in section, and

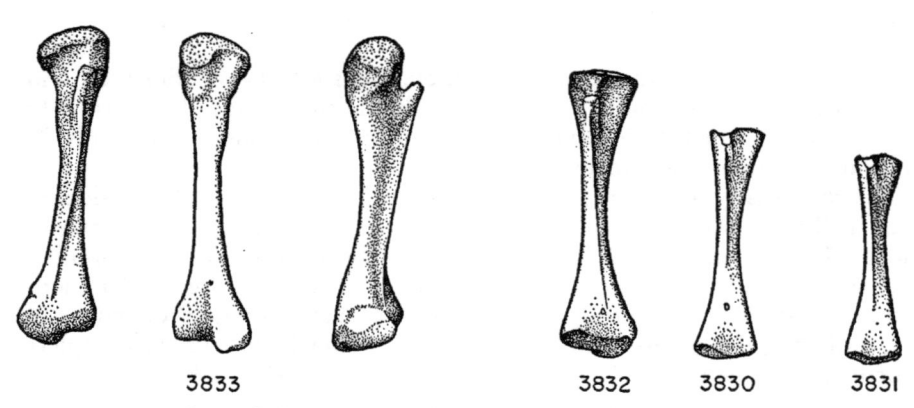

Figure 28. *Cardiocephalus*. Femora from Ft. Sill, Oklahoma. Y.P.M. no. 3833, ventral or adductor surface, dorsal surface and posterior view. Ventral surface of immature femora; Y.P.M. nos. 3832, 3830, 3831. All x 5.

inclined at a slight angle to the shaft. No olecranon process can properly be distinguished—a feature which separates them at once from those of associated *Captorhinus*—but the obscurely triangular section of the shaft below the head serves to distinguish them from the evenly round distal ends of the radii. Distally the ulnae are obliquely compressed. The shaft is a trifle stouter than that of radii of comparable size, a fact which makes their relative scarcity difficult to explain.

Typical specimens are illustrated in figure 27. Lengths of 12 ulnae range from 2.38 mm. to 6.54 mm., the mean being 3.18 ± 0.31 mm.

Excluding a single specimen (no. 3817) which is nearly twice the size of the others (6.54 mm.), the remaining 11 ulnae have a mean length of 2.87 mm., with a standard deviation of 0.36 mm. and coefficient of variability of 12.54. The observed range of this more homogeneous sample is 2.38–3.41 mm. It may be noted that the "oversize" ulna agrees in length with the three aberrantly large radii, and might reasonably belong to the same animal, possibly *Euryodus*.

FEMUR: Numerous small, slender femora from the Fort Sill locality differ from those of *Captorhinus* in an analogous fashion to the humeri, and are judged by their relative abundance to belong to the same kind of animal, namely *Cardiocephalus* (fig. 28). The femur has a straight head, obliquely oval in well formed specimens, a moderately deep, short adductor fossa, and a prominent internal trochanter which projects proximally beneath the head from the down-turned anterior margin of the fossa. A straight adductor crest runs from the base of the internal trochanter to near the distal end of the bone ending at the center of the ventral surface just above the tibial condyles. A small nutrient foramen pierces the popliteal area near the end of the ridge. No fourth trochanter occurs. The condyles are slightly asymmetric as in most early vertebrates, the posterior extending farther distally than the anterior. The fibular condyle is minute.

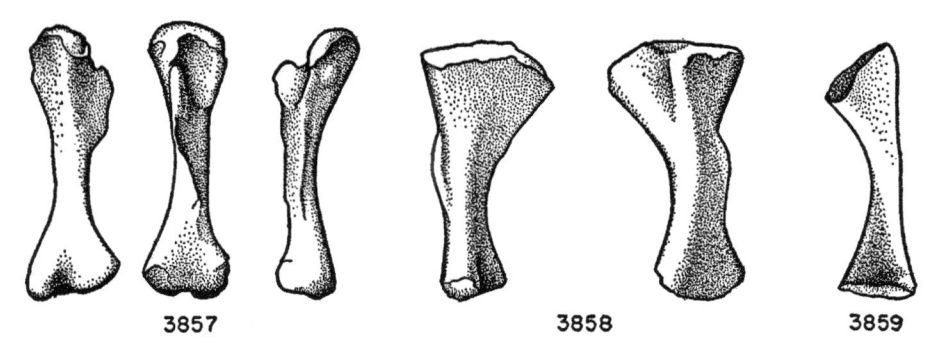

Figure 29. *Captorhinus* sp. from Ft. Sill, Oklahoma. Femur, Y.P.M. no. 3857. Dorsal, ventral, and posterior views. Tibia Y.P.M. no. 3858, preaxial and postaxial views. Fibula Y.P.M. no. 3859, medial surface. x 2.

Cardiocephalus femora differ from those of the associated *Captorhinus* (fig. 29) in greater slenderness and more delicate build, smaller size, and less expansion of the distal end. The adductor ridge is straight and extends far distally whereas in *Captorhinus* it is irregular, terminates well above the distal end, and may have some development of a fourth trochanter for the caudifemoralis muscle.

Measurements are given in table 7. Note that the fully adult specimens with ossified articular surfaces form a compact group with average length 9½ mm.

TIBIA: Numerous small tibiae which differ from those of *Captorhinus* in smaller size, much weaker cnemial crest, and somewhat more pronounced expansion of the distal end, are referred to *Cardiocephalus;* they almost certainly pertain to the animal whose femora have just been described, and occur in comparable abundance. The dorsal surface is slightly convex, the ventral and lateral edges strongly concave. A fine ridge runs the entire length of the lateral edge, doubtless showing attachment of the interosseous membrane. A broad head bears confluent and indistinguishable condyles. Only the larger specimens show any trace of a weak cnemial crest, lateral to which lies a broad shallow sinus (groove for extensor digitorum communis). At the distal end of mature bones two articular facets meet at an obtuse angle, both directed somewhat posteriorly. Immature bones lack articular surfaces and have flat or concave, unfinished ends.

Form and proportions of the tibiae are shown in figure 30. Length varies from 3.12 to 5.51 mm., the mean of 151 specimens being 4.02 mm.

FIBULA: *Cardiocephalus* fibulae (fig. 30) are rather markedly bowed outward, the sharp lateral margin usually convex, the rounded inner surface always strongly concave. The shaft is flattened throughout its length, and twisted slightly, planes of the proximal and distal expansions forming a slight angle. Aside from their small size, they display few features to separate them from the somewhat more slender (relatively) fibulae of *Captorhinus*.

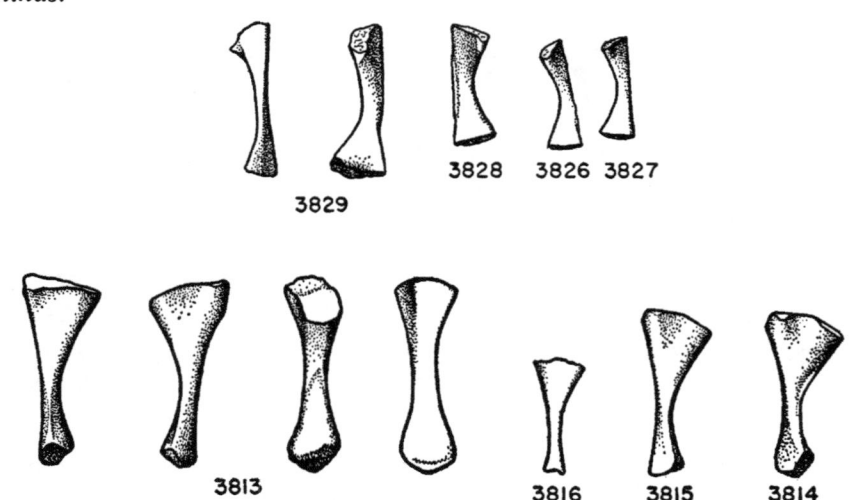

Figure 30. *Cardiocephalus*. Ft. Sill, Oklahoma. Tibiae, Y.P.M. no. 3813, preaxial, postaxial, flexor, and extensor views. Postaxial views of Y.P.M. no. 3816, 3815 and 3814. Fibulae, Y.P.M. no. 3829, lateral and ventral surfaces. Ventral views of Y.P.M. nos. 3828, 3826, and 3827. All x 5.

Forty-four fibulae were recovered. Measurements of complete specimens are given in table 7.

FOOT BONES: Great numbers of minute metapodials and phalanges with imperfect terminations are present, but there is no way of associating these or determining which elements were ossified in *Cardiocephalus*.

DISCUSSION OF MEASUREMENTS OF LIMB BONES: Measurements of several hundred limb bones are summarized in table 7. Probably more significance should be attached to the maximum size (the larger figure of the "Observed range") than to the mean and its dispersion, because immature amphibian bones cannot readily be segregated from those of adults, and have been included in the measured sample. The minimum observed size is of no particular significance; for almost every element, still smaller bones were observed in sorting the material, but these are so fragile that they had been broken either in preparation or by movements within the matrix during entombment, so they are unsuitable for measurement.

The shape of the size frequency distribution suggests an approach to the

TABLE 7
Measurements in millimeters of limb bones

	No. of specimens	Observed range	Mean	Standard deviation	Coefficient of variability
Humerus					
length	40	4.21–8.64	6.23 ± 0.17	1.07 ± 0.12	17.24 ± 1.93
width head	41	1.18–2.18	1.69 ± 0.03	0.22 ± 0.02	13.07 ± 1.44
width distal end	38	1.10–2.64	1.94 ± 0.06	0.34 ± 0.04	17.40 ± 2.00
Radius					
length	25	2.53–4.63	3.36 ± 0.09	0.45 ± 0.06	13.54 ± 1.92
Ulna					
length	11	2.38–3.41	2.87 ± 0.11	0.36 ± 0.08	12.59 ± 2.68
Femur					
length all	141	5.17–10.16	7.35 ± 0.08	0.94 ± 0.06	12.80 ± 0.76
length (with ossified condyles)	14	8.35–10.16	9.48 ± 0.11	0.42 ± 0.08	4.39 ± 0.83
Tibia					
length	151	3.12–5.51	4.02 ± 0.04	0.53 ± 0.03	13.13 ± 0.76
Fibula					
length	35	2.88–5.06	3.72 ± 0.09	0.55 ± 0.07	14.80 ± 1.77

age distribution of the *Cardiocephalus* population of the fissures. Histograms for femora, tibiae, and fibula (fig. 32) all show a moderate positive skewness (+0.73 and +0.74 for femur and tibia respectively, +0.22 for the much smaller fibula sample). Assuming that size varies directly with

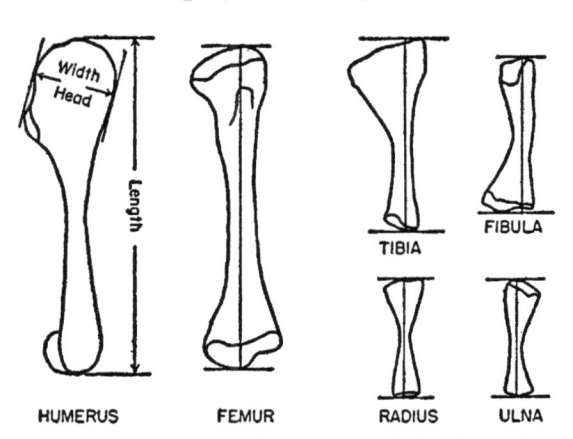

Figure 31. Index to measurements of *Cardiocephalus* limb bones.

age, the bulk of the population was immature. Perhaps the few femora and humeri with "finished" terminations represent the adults; for the femora these are only one-tenth of the entire group. The possibility, however, that some of these larger bones belong to *Euryodus* rather than to *Cardiocephalus* makes further speculation on population structure unprofitable.

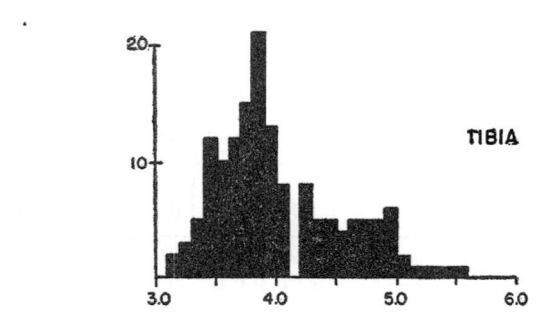

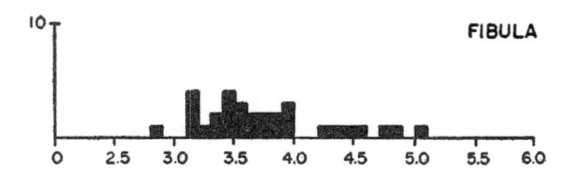

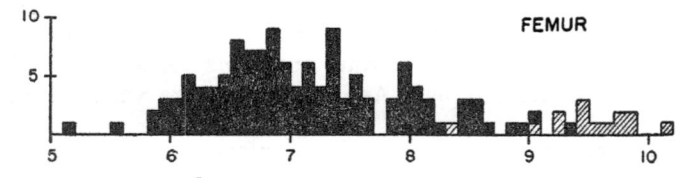

Figure 32. Histograms of variation in length of microsaur limb bones from fissures north of Ft. Sill, Oklahoma. Oblique shading indicates femora with finished ends.

RELATIONSHIPS OF THE MICROSAURS

Romer (1950, p. 645–651) has summarized the history of opinion on microsaur relationships and advanced reasons for considering them to be Amphibia, not near the ancestry of reptiles, and possibly close to the ancestry of the urodeles. Structure of the auditory region of *Cardiocephalus* strongly supports his interpretation. The stapes is utterly unlike that of such primitive reptiles as *Captorhinus, Dimetrodon,* or *Ophiacodon;* instead it resembles the urodeles in its large footplate and minute, imperforate columellar process. The large fenestra ovalis low on the posterolateral corner of the braincase has quite different relationships from that of reptiles. Evidence of the single scapulocoracoid and pelvic ossifications, and absence of an entepicondylar foramen in the humerus may be added to Romer's list of non-reptilian features.

Relationship of the microsaurs to the urodeles rather than to the gymnophionans is less certain. The general form of the stapes of *Cardiocephalus* resembles about equally that of both existing Orders. At least in *Ichthyophis* among the Gymnophiona, the columella is perforated by the stapedial artery (Noble, 1931, p. 222; de Beer, 1937, p. 193), and Gadow (1923, p. 85) states that the stapes of *Ichthyophis* is stirrup shaped and perforated. However dried skeletons of *Siphonops* and *Caecilia* show no stapedial foramina. Ossification of the stapes commences in the footplate and generally does not proceed far up the shaft. If a foramen is not formed the stapes does not differ greatly from that of the urodeles in which it is imperforate.

The fenestra ovalis of the Gymnophiona is directed laterally; that of the microsaurs more ventrally than laterally. Many salamanders also have the fenestra ovalis far out on the side of the braincase, but the conservative genus *Ambystoma* is more like the microsaurs.

Cardiocephalus differs from all living lepospondylous amphibians and approaches the aistopod *Phlegethontia* in the large posteroventral fenestra ovalis and relatively large size of the stapes in comparison to the skull as a whole. Among Paleozoic Amphibia, *Lysorophus* more closely resembles recent forms, especially the gymnophionans, in its auditory structures.

Microsaurs, aistopods, and *Lysorophus* all resemble both salamanders and caecilians in the position of the forwardly sloping quadrate, situated not far in front of the occiput and immediately adjacent to the stapes and fenestra ovalis. *Lysorophus,* with is squamosal forked for reception of the supratemporal and small, ventral quadrate, closely resembles such urodeles as *Salamandra* or *Necturus. Amphiuma* and especially *Ambystoma,* in which the quadrate is farther back and more vertical, more closely resemble *Cardiocephalus*. The extreme forward position of the jaws in the gymnophionans gives their suspensorium a forward position superficially similar to that of *Lysorophus* or the aistopods, but this may be attributed to convergence; moreover *Necturus* has its suspensorium as far forward although different in detail.

Neither the adult structures nor the embryonic development of this region affords features surely indicative of relationship. On the whole the lepospondyls are quite similar to one another in the relationships of quadrate, squamosal, and braincase.

As Romer (1950, p. 636) has pointed out, the microsaur palate is primitive in having a movable basal articulation, a narrow cultriform process of the parasphenoid and narrow interpterygoid vacuities. The parasphenoid tapers anteriorly, and is perhaps intermediate between the narrow type found in such primitive labyrinthodonts as *Palaeogyrinus* or *Edops* and the uniformly broad condition which characterizes the urodeles. The posterior broadening of the microsaur parasphenoid is related to the form of the braincase and is similar to that of advanced labyrinthodonts. No feature of the palate points conclusively in the urodele direction, though the structure is such that the salamander palate could be derived from it by broadening of the anterior part of the parasphenoid and reduction of pterygoids and lateral palatal elements.

Gymnophionans have the most primitive palate of any living amphibian, in that all bones save the ectopterygoid are retained. Special resemblances to the microsaurs include the very wide posterior part of the parasphenoid, which reaches the fenestrae ovales, and the wedge-shaped anterior portion of that bone. Basipterygoid processes, however, are less developed than in microsaurs, owing perhaps to extensive fusion of palate with braincase in adaptation to burrowing habits. Resemblances between microsaur and gymnophionan palates are not convincing evidence of relationship; but no features would bar descent of the latter from the former, and the resemblances are considerably greater than those between microsaurs and the greatly modified palates of urodeles.

In contrast, the greatly reduced palate of *Lysorophus* affords little basis for deriving that of the Gymnophiona.

In their completely roofed skull, *Cardiocephalus* and other typical microsaurs are far more primitive than any salamanders. Urodele skulls are diverse in structure; all have lost many of the primitive tetrapod elements, but different elements, particularly of the palate, are present in various modern genera. Any of these conditions could be derived from the skull of *Cardiocephalus*, or some other microsaur, by appropriate reduction of ossification.

At first sight the elongate postorbital region of *Cardiocephalus* appears to differ markedly from the salamander skull, but this results from the deceptive enlargement of the orbitotemporal fenestra in the urodele skull roof. Salamander parietals are elongate, and while the postparietals have been lost, the proportions of the skull with respect to the eyes remain much the same in the modern as in the ancient Order.

In contrast to urodeles, gymnophionans have solidly roofed skulls. Opinions have differed as to whether this is primitive or a secondary feature associated with the burrowing adaptations of these animals. Marcus, Stimmelmayr, and Porsch (1935 [quoted by de Beer, 1937, p. 196–197]) have shown that the adult skull roof of *Hypogeophis* is the product of extensive fusions of ossification centers present in the larva. Centers corresponding to all primitive roofing bones except the supratemporal, post frontal, and quadratojugal have been recognized. The prominence of the supratemporal in the microsaur skull might lead one to expect that it would be retained, as a center of ossification at least, if gymnophionans were descendants of microsaurs. However, if salamanders were derived from microsaurs the loss of ossification was far more extensive, and the final skull pattern is far more modified than that of caecilians.

Microsaur vertebrae are among the most primitive of the lepospondyl group, their chief specializations being reduction or loss of the neural spine and the wide low neural arches with widely separate zygapophyses. In these features they resemble some salamanders, all gymnophionans, and aistopods, a resemblance probably in part adaptive for small, relatively weak, but supple burrowing animals. The primitive, anterior position of the transverse process, which resembles that of reptiles, is shared only with the Gymnophiona, and may well be significant of closer relationship between these Orders than with the other Lepospondyli. On the other hand the gymnarthrid atlas (p. 43–44, fig. 16A) strongly resembles that of urodeles and differs from that of caecilians in its relatively greater height and particularly in its well developed odontoid process. Articular cotyli of a gymnophionan atlas closely approach the midline to receive the almost confluent occipital condyles. There is no intercondylar pit in the basioccipital to accommodate a projecting odontoid, and no odontoid process is present at least in *Siphonops*. Although the form of the gymnarthrid atlas certainly is more suggestive of urodele than caecilian relationships, it is possible that the atlanto-occipital joint of the latter is part of their extreme fossorial adaptation. If so, the rib articulation may well be more conservative and phylogenetically significant.

It seems worth pointing out that salamander vertebrae differ from those of microsaurs and gymnophionans in the position and structure of the transverse process, which arises near the middle of the vertebra and in general bears two articular facets for the rib. In this they resemble the vertebrae of nectridians and to a lesser extent aistopods. Moreover, caudal vertebrae of salamanders develop auxiliary articulations between neural and haemal spines of adjacent vertebrae, similar to the Nectridia. In view of the equivocal nature of the skull patterns, it seems reasonable to place considerable emphasis upon these vertebral structures in assessing the relationships of these Orders.

Microsaur vertebrae, and those of the lysorophids, are unique among amphibians and resemble those of reptiles in the separate ossification of the hourglass-shaped centra and neural arches. Watson (1929) and Steen

(1938, p. 272) separated these families as a Subclass Adelospondyli in contrast to the remaining Lepospondyli in which the entire vertebra is a single ossification. Much skepticism has been expressed, at least orally, about the value of the presence or absence of the neurocentral suture as a feature of fundamental taxonomic importance. In reptiles and mammals neural arches are characteristically separate ossifications united to the centrum by sutures in the young, but the two frequently become indistinguishably fused in mature individuals. Nonetheless such vertebrae contrast strikingly with those of salamanders, for example, in which no trace of separate ossifications or a suture may be found even in early developmental stages. Romer (1945, p. 158, 591–592) included the adelospondylous amphibians within the Lepospondyli as a single Order, Microsauria.

It must be pointed out, therefore, that vertebrae of the Gymnophiona are similar to those of salamanders and unlike those of microsaurs in the absence of a neurocentral suture at any stage of development. This evidence opposes the implications of transverse processes and rib articulations as to possible relationships between caecilians and microsaurs. If the character of the neurocentral suture is relied upon, microsaurs and lysorophids are related to one another and remote from the remaining lepospondyls; moreover they cannot be regarded as ancestral to either the living urodeles or caecilians, with which they have many features in common. Is this detail of vertebral development more fundamental than the resemblances in form of centrum, form of stapes, and its relationship to the quadrate, all of which tend to unite these families with the remaining lepospondyls? Skull patterns point to descent of both lepospondylous and apsidospondylous Amphibia from a common crossopterygian ancestor. Hence the lepospondylous vertebrae must ultimately have been derived from a rhachitomous type involving several ossifications. Is it not possible, then, that strictly lepospondylous modern gymnophionan vertebrae could be derived from the primitive microsaur vertebrae in which neural arch and centrum still ossified separately? Unfortunately the lack of intermediate fossils renders any conclusion on this point highly speculative.

On the basis of present knowledge, one must take one of two positions on lepospondyl phylogeny:

(1) Watson's view, that microsaurs and lysorophids constitute a distinct group, the Adelospondyli, of fundamentally different vertebral organization than the Lepospondyli.

(2) Regard the presence or absence of a neurocentral suture as unreliable for classification in opposition to similarities in the auditory region.

The latter viewpoint appears most probable to us. Instead, the position of the rib articulation is suggested as the most promising basis for supraordinal grouping. On this basis, urodeles and aistopods are related to nectridians whereas gymnophionans and lysorophids are closer to the microsaurs.

The primitive girdles and limb bones of *Cardiocephalus* and other microsaurs do not differ from those expected in any primitive tetrapod, and show no special features suggestive of any modern Order. Perhaps the relatively small size of the front limb is worth noting, as a resemblance to salamanders; the limbless caecilians of course must ultimately have been derived from more normal limb-bearing ancestors, and the reduced forelimb of microsaurs may equally well be the beginning of the trend which led to loss of legs.

That microsaurs commonly possessed scales has been known since the work of Fritsch and Credner; Romer (1950, p. 633) maintains that their pattern of radiate striations is unique for tetrapods; they appear similar to the "radii" of cycloid fish scales. Several living genera of Gymnophiona have bony scales embedded in the skin; the ornamentation of these is concentric, however, and similar to the "circuli" of cycloid fish scales. Accordingly no special resemblance may be said to exist between the two scale types. However circuli are essentially growth increments of the scale (not, of course, to be confused with annual growth lines). By suppression of the radiate pattern, a microsaur scale would be transformed into a caecilian scale. Fairly closely related groups of fishes often differ considerably in development of radii and other pattern elements. Hence the difference in scale pattern should not be regarded as evidence against the relationship.

In summary, the following phylogenetic hypotheses may be tentatively advanced, in full realization of the inadequacy of data to fully substantiate them:

The Orders of Lepospondyli may be divided into two groups on the basis of vertebral structure.

1. Vertebrae of the Microsauria and Gymnophiona have transverse processes at the anterior end of the neural arch, similar in general form to those of reptiles, and articulating with the tuberculum of the rib. The rib capitulum in these Orders articulates intervertebrally or with a parapophysis at the anterior edge of the centrum.

Lysorophus and *Megamolgophis* have vertebrae similar to microsaurs, although the transverse process is farther back. Probably they are related to this group.

2. Vertebrae of the Nectridia, Aistopoda, and Urodela have transverse processes arising from near the middle of the centrum, usually bearing two articular facets. In salamanders the ventral rib head is homologous with the tuberculum and a secondary more dorsal articulation has developed, the capitulum being lost. Whether these conditions had already been attained by the related Paleozoic Orders is unknown.

Microsauria are the most primitive and conservative of the Lepospondyli, having a normal, though slightly elongated body form, relatively well developed limbs, and a skull of normal proportions.

Gymnophiona are specialized descendants of the microsaurs which have lost the limbs, girdles, and tail, and have undergone extensive fusion of skull bones in adaptation to burrowing life. They retain cycloid scales somewhat similar to those of microsaurs embedded in the skin.

Nectridia are the most conservative of the Orders which have intervertebral transverse processes and rib articulations. *Sauropleura* and *Ptyonius* retain fairly primitive skull patterns; other genera become specialized in the development of horns. Nectridian vertebrae are specialized in the high interlocking neural and haemal spines of the caudal vertebrae, but possibly are ancestral to the Urodela, in which vestiges of these structures occur.

Urodela are highly specialized in the reduction of the roofing bones of the skull, and in part show compensation in the enlargement of remaining elements, particularly the parasphenoid. Their vertebral structure is derivable from that of the Nectridia.

Aistopoda are a highly specialized limbless group of great antiquity. Skull structure and general form of the vertebral column recall the burrowing Gymnophiona and *Lysorophus,* but the transverse process is situated as in nectridians and urodeles. Aistopods may provisionally be regarded as an early offshoot of the nectridian stock.

The accompanying phylogenetic diagram represents the view here advanced (fig. 33).

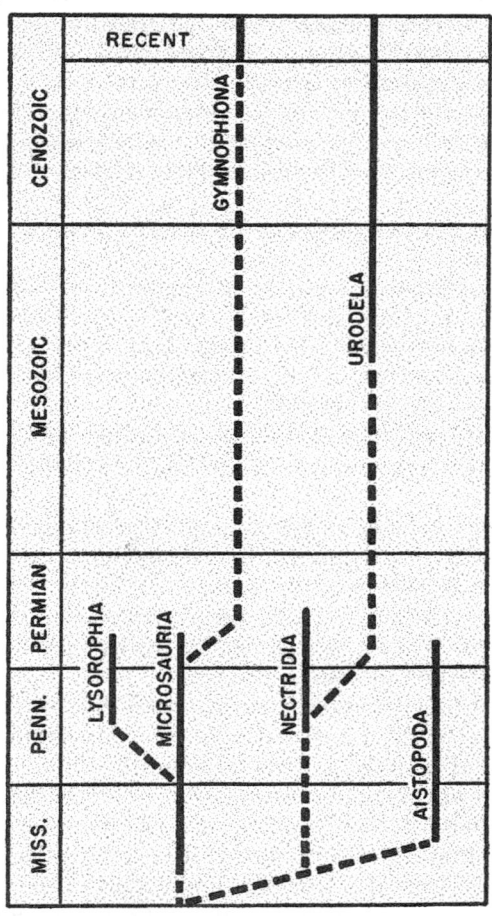

Figure 33. Suggested phyletic relationships of lepospondyl Orders based upon vertebral form.

RESTORATION

Cardiocephalus belongs to a slightly aberrant group of microsaurs sometimes referred to as the gymnarthrids (from a synonymous generic name). They are characterized by a completely roofed skull, lacking an otic notch and having relatively short preorbital and long postorbital regions, a large supratemporal, and well developed postparietals. The neck is elongate and reptile-like (in contrast to most amphibians). No specimens sufficiently complete to show the length of the vertebral column have been discovered, but the articulated skeleton (O.U. no. 1034) described earlier suggests an elongate salamander-like body similar to the "typical" microsaurs *Microbrachis* and *Hyloplesion*. Limbs are feeble in comparison to body size, and the fore limbs are appreciably smaller than the rear. Osseous scales are not known, but may have been present as in other microsaurs. In life they may have closely approached modern salamanders in both appearance and habits (Frontispiece).

Cardiocephalus has short and stout though pointed teeth adapted for cutting food by strong shearing bites and possibly somewhat useful for crushing. Conceivably it could have been herbivorous, feeding on roots or underground stems, but its dentition seems equally well suited for feeding on worms, insects, and other small animals which dwell in the ground. It may be regarded as moderately predaceous, with a relatively strong bite for an animal its size, adapted to killing its prey by powerful biting or crushing action of its jaws rather than by piercing with elongate, caniniform teeth.

The slightly larger *Euryodus* with its enlarged, pointed piercing tooth, seems more surely predaceous, and may have fed on larger prey, perhaps even small vertebrates. The crushing dentition of *Pantylus,* representing another adaptation within the family, might have been an adaptation to mollusk feeding.

Gymnarthrids resemble the living gymnophionans in their solidly roofed skulls, well ossified braincases, forwardly directed quadrate and short mouth, well developed retroarticular process of the lower jaw, and particularly in their vertebrae which lack neural spines and have low, horizontal zygapophyses. Similar vertebrae and skull modifications are characteristic of all tetrapods adapted to a burrowing, subterranean mode of life, and their presence in *Cardiocephalus* strongly suggests fossorial habits. The diminutive forelimbs of this animal likewise suggest an early stage of reduction, which in more specialized stages has produced the anomaly of completely limbless tetrapods.

Some of these structures are also characteristic of snakes and such lizards as the anguids whose locomotion is predominantly by body movements rather than the legs. The well developed orbit of the gymnarthrids might be considered more suggestive of such an adaptation for life on the surface of the ground than for subterranean life. But the resemblance of the head and particularly the jaw to burrowing forms, especially caecelians, is more impressive. Likewise the occurrence of these animals in a cave deposit suggests a subterranean habitat. Aside from the Oklahoma fissure deposits, gymnarthrids are known principally from the "*Lysorophus* pockets" of the Clear Fork in north Texas where they are associated with remains of the highly specialized limbless amphibian *Lysorophus*. Olson (1939, p. 396) interprets these deposits as those of pools along intermittant streams. If the gymnarthrids were indeed fossorial, moist mud of such pool bottoms would be a favorable place for their burrows.

REFERENCES CITED

Beer, G. R. de, 1937, The development of the vertebrate skull: Oxford, Clarendon Press, xxiv, 552 p., 143 pls.

Broili, Ferdinand, 1904, Permische Stegocephalen und Reptilien aus Texas: Palaeontographica, v. 51, 120 p., 13 pls. 5 figs.

——, 1913, Über zwei Stegocephalenreste aus dem texanischen Perm.: Neues. Jahrb. f. Min., Geol., und Paleontology, Jahrg. 1913, Bd. I, p. 96–100, pl. IX.

Broom, Robert, 1910, A comparison of the Permian reptiles of North America with those of South Africa: Am. Mus. Nat. Hist. Bull., v. 28, p. 197–234, 20 figs.

——, 1930, The origin of the human skeleton; an introduction to human osteology: London, Witherby, 164 p., 46 figs. 2 folded charts.

Case, E. C., 1907, Revision of the Pelycosauria of North America: Carnegie Inst. Washington Pub. 55, 176 p., 35 pls., 75 figs.

——, 1910, New or little known reptiles and amphibians from the Permian (?) of Texas: Am. Mus. Nat. Hist. Bull., v. 28, p. 163–181, 10 figs.

——, 1911a, A revision of the cotylosauria of North America: Carnegie Inst. Washington Pub. 145, 121 p., 14 pls., 52 figs.

——, 1911b, Revision of the Amphibia and Pisces of the Permian of North America: Carnegie Inst. Washington Pub. 146, 179 p., 32 pls., 56 figs.

——, 1929, Description of a nearly complete skeleton of *Ostodolepis brevispinatus* Williston: Univ. Michigan, Museum of Paleontology Contr., v. 3, no. 5, p. 81–107, 8 pls., 12 figs.

——, 1946, A census of the determinable genera of the Stegocephalia: Am. Philos. Soc. Trans., v. 35, p. 325–420.

Cope, E. D., 1878, Descriptions of extinct Batrachia and Reptilia from the Permian formation of Texas: Am. Philos. Soc. Proc., v. 17, p. 505–530.

——, 1881 [1882 imprint?], On some new Batrachia and Reptilia from the Permian beds of Texas: U.S. Geol. and Geog. Survey Terr. Bull., v. 6, p. 79–82.

——, 1883, Fourth contribution to the history of the Permian formation of Texas: Am. Philos. Soc. Proc., v. 20, p. 628–636.

——, 1895, The reptilian Order Cotylosauria: Am. Philos. Soc. Proc., v. 34, p. 436–457, pls. 7–9.

——, 1896, Second contribution to the history of the cotylosauria: Am. Philos. Soc. Proc., v. 35, p. 122–139, pls. 7–10.

Cummins, W. F., 1908, The localities and horizons of Permian vertebrate fossils in Texas: Jour. Geol., v. 16, p. 737–745.

Fritsch, Anton, 1879–1883, Fauna der Gaskohle und der Kalksteine der Permformation Böhmens: Bd. 1, Prag. 182 p., 48 pls., 116 figs.

Gadow, Hans, 1923, Amphibia and reptiles: The Cambridge Natural History, v. 8, xiii, 668 p., London, Macmillan.

Gregory, J. T., 1948, A new limbless vertebrate from the Pennsylvanian of Mazon Creek, Illinois: Am. Jour. Sci., v. 246, p. 636–663.

HAY, O. P., 1929, Second bibliography and catalogue of the fossil Vertebrata of North America: Carnegie Inst. Washington Pub. 390, v. 1, viii, 916 p.

HUENE, FRIEDRICH VON, 1913, The skull elements of the Permian Tetrapoda in the American Museum of Natural History, New York: Am. Mus. Nat. Hist. Bull., v. 32, p. 315–386, 57 figs.

MARCUS, H., STIMMELMAYR, E., and PORSCH, G., 1935, Die Ossifikation des *Hypogeophis*-schädels: Morph. Jahrb., v. 76, p. 373–420.

MEHL, M. G., 1912, *Pantylus cordatus* Cope: Jour. Geol., v. 20, p. 21–27, 2 figs.

NOBLE, G. K., 1931, Biology of the Amphibia: New York, McGraw-Hill, xiii, 577 p.

OLSON, E. C., 1939, The fauna of the *Lysorophus* pockets in the Clear Fork Permian, Baylor County, Texas: Jour. Geol., v. 47, p. 389–397, 2 figs., 1 pl.

———, 1954, Fauna of the Vale and Choza; 9. Captorhinomorpha: Chicago Nat. Hist. Mus., Fieldiana, Geol., v. 10, no. 19, p. 211–218, figs. 85, 86.

PRICE, L. I., 1935, Notes on the brain case of *Captorhinus*: Boston Soc. Nat. Hist. Proc., v. 40, no. 7, p. 377–386, pls. 6–9.

———, 1940, Autotomy of the tail in Permian reptiles: Copeia, 1940, p. 119–120, 1 fig.

ROMER, A. S., 1928, Vertebrate faunal horizons in the Texas Permo-Carboniferous redbeds: Texas Univ. Bull. 2801, p. 67–108, 1 fig.

———, 1935, Early history of Texas redbeds vertebrates: Geol. Soc. Am. Bull., v. 46, p. 1597–1658.

———, 1945, Vertebrate Paleontology: Rev. ed., Univ. Chicago Press, viii, 687 p., 377 figs.

———, 1947, Review of the Labyrinthodontia: Harvard Univ., Mus. Comp. Zool. Bull., v. 99, no. 1, 368 p.

———, 1950, The nature and relationships of the Paleozoic microsaurs: Am. Jour. Sci., v. 248, p. 628–654, 4 figs.

SOLLAS, W. J., 1920, On the structure of *Lysorophus*, as exposed by serial sections: Royal Soc. London Philos. Trans. (B) 109, p. 481–527.

STEEN, M. C., 1938, On the fossil Amphibia from the Gas Coal of Nýřany and other deposits in Czechoslovakia: Zool. Soc. London Proc., v. 108, ser. B., p. 205–283, 7 pls., 47 figs.

SUSHKIN, P. P., 1927, On the modifications of the mandibular and hyoid arches and their relations to the brain case in the early Tetrapoda: Palaeontologisches Zeitschrift, Bd. 8, p. 263–321, 39 figs.

WATSON, D. M. S., 1929, The Carboniferous Amphibia of Scotland: Pal. Hungarica, v. 1, p. 219–252.

WILLISTON, S. W., 1909, New or little known Permian vertebrates. *Pariotichus*: Biol. Bull., v. 17, p. 241–255, 6 figs.

———, 1911, American Permian Vertebrates: Univ. Chicago Press, 145 p., 38 pls., 32 figs.

———, 1913, *Ostodolepis brevispinatus*, a new reptile from the Permian of Texas: Jour. Geol., v. 21, p. 363–366, 2 figs.

———, 1914, The osteology of some American Permian vertebrates: Univ. Chicago, Walker Mus. Contr., v. 1, no. 8, p. 107–162, 19 figs.

———, 1916a, The osteology of some American Permian vertebrates, II: Univ. Chicago, Walker Mus. Contr., v. 1, no. 9, p. 165–192, figs. 20–37.

———, 1916b, Synopsis of the American Permocarboniferous Tetrapoda: Univ. Chicago, Walker Mus. Contr., v. 1, no. 9, p. 193–236.

———, 1925, The osteology of the reptiles: Cambridge, Harvard Univ. Press, xii, 300 p., 191 figs.
WILSON, J. A., 1951, Taxonomic position of *Pantylus* (abstract): Geol. Soc. Am. Bull., v. 62, p. 1490–1491.
ZANGERL, RAINER, 1944, Contributions to the osteology of the skull of the Amphisbaenidae: Am. Midland Nat., v. 31, p. 417–454.

INDEX

Numbers in bold face type indicate page on which the reference is illustrated.

Adaptation,
  dentition, 40, 69
  fossorial, 69
Adelospondyli, 37, 65, 66
Admiral formation, 6
Aistopod, Ft. Sill, Okla., 3
Aistopoda, relationships, 63, 66, 67
Ambystoma, 36, 63
Amphiuma, 63
Apache, Okla., 1, 2
Araeoscelis, 48
Arroyo fauna, 3
Arroyo formation, 9, 26, 31, 32
Articular region, 16, 23
Atlas, 25, 30, 43, 65
Auditory region, 35–37, **36**, 63

Basicranial region, 22, 28, 41
Basisphenoid articulation, 21, 22, 42
Basioccipital, 41, **42**
Beaver Creek, Texas, 9
Big Wichita River, Texas, 30, 31
Boll, Jacob, 6
Bolosaurus rapidens, 6
Bolosaurus vertebrae, 48
Braincase, 16, 22, 42, 43
Brushy Creek, Texas, 26

Caecilians, see Gymnophiona
Camp, C. L., notes, 9
Captorhinidae, 5
Captorhinus, Ft. Sill, Okla., 3
Captorhinus aduncus, 8
Captorhinus aguti, 8
Captorhinus angusticeps, 8
Captorhinus isolomus, 8
Cardiocephalus, 3, 9, 33, 34
  distinctions from *Captorhinus*, 25, 47, 55, 59
  distinctions from *Euryodus*, 26
Cardiocephalus sternbergi, 9–14, 38
Cardiocephalus cf. *sternbergi*, 14–26, 41–62
  articulated skeleton, **24**
Carotid, internal, canal, 22
Case, E. C., 5
Chicago Museum of Natural History, 2
Chicago, University of,
  Walker Museum, 11

Chilonyx rapidens, 6
Clear Fork group, 9, 26, 32
Clyde formation, 31
Coffee Creek, Texas, 9, 31
Cope, E. D., 5
Coronoid, 23, 26, **30**
Cryptobranchus, 41, 43

Dentition, 16–19, **20**, 28, 39, 40
  adaptation, 40, 69
  coronoid, 23, 24, 30, 40
  gymnarthrids, 38
  labyrinthine structure, 39
  tooth counts, 18, 29
  tooth implantation, 38
  tooth replacement, 38
Dolese Brothers' Limestone Quarry, 2

Ectocynodon aguti, 8
Ectocynodon incisivus, 8
Ectocynodon ordinatus, 8
Edops, 64
Euryodus, 22, 26, 33, 34, 42–43
  distinctions from *Cardiocephalus*, 26
  Ft. Sill, Okla., 3, 28
Euryodus primus, 26–30
Exoccipital, 41, **42**

Fauna,
  Ft. Sill locality, 3
Fenestra ovalis, 16, 35
Femur,
  Cardiocephalus, **58**, **61**, **62**
  Captorhinus, **59**
Fibula,
  Cardiocephalus, **60**, **61**
  Captorhinus, **59**
Foot bones, 60
Fort Sill, Oklahoma, 1
  locality, 2, 3, 28, 41, **Frontispiece**
Fulda, Texas, 9

Gymnarthria, 5
Gymnarthridae, 5, 32–34
  key to genera, 33
  stratigraphic distribution, 34
Gymnarthrus, 11
  supposed distinctions from *Cardiocephalus*, 11, 13

Gymnarthrus willoughbyi, 9, 10, 12
Gymnophiona, 36, 64, 65
  relationships of, 63, 66, 67
Goniocara, 30, 34
Goniocara willistoni, 32
Goniocephalus willistoni, 32

Hackberry Creek, Texas, 9
Humerus,
  Captorhinus, 56
  Cardiocephalus, 55, 56, 61
Hyloplesion longicaudatum, 47
Hypogeophis,
  ossification centers of, 65
Hypoglossal foramina, 42, 43

Ichthyophis, 63
Ilium, 50
Ischium, 51
International Rules of Zoological Nomenclature, 32
Isodectes megalops, 2

Jaws,
  Cardiocephalus cf. *sternbergi*, 16, 17, 20, 21, 22
  Euryodus, 23, 30

Kansas University Museum of Natural History, 2
Key to genera, 33

Labidosaurus, Ft. Sill, Okla, 3
Labidosaurus hamatus, 8
Labyrinthodonts, Ft. Sill, Okla., 3
Lacrimal, 19
Lepospondyli, 32, 65, 66, 67, 68
Limb bones,
  Cardiocephalus, 26, 56–60
  measurements, 61
  variation in, 62
Little Wichita River,
  North Fork, 6
Lysorophus, 37, 63, 67, 68

Measurements,
  limb bones, 61
  scapulocoracoid, 52
  skull, 14
  vertebrae, 49
Megamolgophis, 67
Microbrachis, 47
Microsauria, 32, 67
Microsaurs,
  relationships of, 63–68

Mitchell Creek, Texas, 31
Munich,
  Alte Akademie, 9

Necturus, 43, 63
Nectridians, 47, 66, 68

Occipito-atlas region, 41
Odontoid, 43
Oklahoma, University of, 1
Opisthotic, 35
Osteoderms, 10
Ostodolepis, 2, 31, 34, 47
Ostodolepis brevispinatus, 31
Otic notch, 28

Palaeogyrinus, 64
Palate, 16, 19, 21, 36, 64
Pantylidae, 2
Pantylus, 2, 24, 30, 34
Pantylus coicodus, 31
Pantylus cordatus, 30
Pantylus tryptichus, 31
Parasphenoid, 16, 20, 22, 28, 42, 64
Parietal foramen, 9, 11, 13, 15
Pariotichidae, 5, 32
Pariotichus, 5, 33
Pariotichus aduncus, 8
Pariotichus brachyops, 6–8, **6, 7**
Pariotichus hamatus, 8
Pariotichus isolomus, 8
Pariotichus laticeps, 8
Pariotichus megalops, 8
Pelvis, 50, 52, 53, 55
Pelycosaurs, 3
Phlegethontia, 37, 63
Phylogeny, 67, 68
Postcranial bones, 41
Proatlas, 44
Ptyonius, 67
Putnam formation, 6, 31

Quadrate, 63

Radius, 57, 61
Relationships,
  aistopods, 63, 66, 67
  gymnophionans, 63, 66
  Lysorophus, 67
  microsaurs, 63, 66, 68
  nectridians, 66
  phyletic, 68
  urodeles, 66, 67
Restoration of skull, 69
  Cardiocephalus sternbergi, **10**
  Pariotichus brachyops, **7**

INDEX

Ribs, 49, **51**
"Richard's Spur," 3
Romer, A. S., 1, 63
   Vertebrate Paleontology, 1, 5
Romeria, 3

Sacral rib, 51
Salamanders, *see* Urodeles
Salamandra, 63
Sauropleura, 67
Scapulocoracoid, **50**, **51**, 52
Scotland, Texas, 31
Septomaxillary, 19, **20**, 28
Siphonops, 63, 65
Skull,
   Cardiocephalus sternbergi, **10**, 12
   Cardiocephalus cf. *sternbergi*, 14, **15**, 19–26, **20**, **25**
   Euryodus primus, **27**
   Gymnarthrus willoughbyi, 12
   Pariotichus brachyops, **6**, **7**
Slime canals,
   Gymnarthrus, 13
Splenial, 23
Stapes,
   aistopods, 36
   Caecilia, 63
   Cardiocephalus, 16, **35**, **36**, 63
   gymnophionans, 36, 63
   Ichthyophis, **37**, 63
   Lysorophus, 36

Phlegethontia, **37**, 63
Siphonops, 63
   urodeles, 36
Stegocephalia, 11

Tibia, 59, **59**, **60**, 61, 62
Tomicosaurus, 48

Ulna,
   Cardiocephalus, **57**, **58**, 61
   Captorhinus, **56**
   Euryodus, **57**
Urodeles, 47, 66, 68
   relationship to microsaurs, 63, 66–68

Varanosaurus, 45
Vertebra,
   atlas, 25, 30, 43, 65
   Cardiocephalus, 25, 44–49
   Captorhinus, **45**, **46**, 48
   Euryodus, **30**
   Gymnophiona, **47**, **65**
   measurements, 49
   microsaur, **65**
   salamander, **65**

West Coffee Creek, Texas, 31
Wichita group, 6, 31
Wichita Mountains, 2

Yale Peabody Museum, 1

Lightning Source UK Ltd.
Milton Keynes UK
UKHW020916220119
335965UK00013B/1795/P